ELEMENTARY
REAL ANALYSIS

ELEMENTARY REAL ANALYSIS

Kenneth W. Anderson

Associate Professor of Mathematics
State University of New York
at Binghamton

Dick Wick Hall

Professor of Mathematics
State University of New York
at Binghamton

McGRAW-HILL BOOK COMPANY

New York
St. Louis
San Francisco
Düsseldorf
Johannesburg
Kuala Lumpur
London
Mexico
Montreal
New Delhi
Panama
Rio de Janeiro
Singapore
Sydney
Toronto

ELEMENTARY REAL ANALYSIS

Library of Congress Catalog Card Number 74-167489

07-001620-8

1 2 3 4 5 6 7 8 9 0   BPBP   7 9 8 7 6 5 4 3 2

This book was set in Caledonia by Progressive Typographers, and printed and bound by The Book Press. The designer was Anne Canevari; the drawings were done by John Cordes, J. & R. Technical Services, Inc. The editors were Howard S. Aksen, Lee W. Peterson, and Andrea Stryker-Rodda. John A. Sabella supervised production.

CONTENTS

PREFACE

Many recent books have essentially the same objective; bridging the gap between calculus and advanced calculus. The normal undergraduate college curriculum allows at most one semester to accomplish this worthwhile objective. Any rigorous approach to the basic theorems of analysis requires a preliminary discussion of some aspects of logic, set theory, and algebra. We have attempted to meet this requirement with an adequate but minimal treatment of these topics in order to attain our main objective within the required time.

In addition to these preliminary concepts, Part One contains such topics as connected and separated sets, cluster points, the Heine-Borel and Bolzano-Weierstrass Theorems. Part Two introduces the concepts of mappings, difference quotients, continuity, sequences, including the Cauchy criterion for convergence, infinite series, and sets of measure zero. The final part contains the most elementary theorems of real analysis, including a discussion of open and closed sets, compactness, the Cantor product theorem, differentiable functions and their properties, and Taylor's formula with remainder. The last chapter, on The Cantor Set and the Cantor Function, is intended for supplementary reading by students in preparation for their next course in analysis.

This text has been used over the past few years in a one-semester, four-hour Introduction to Analysis course at the State University of New York at Binghamton, as a prerequisite for both advanced calculus and topology. We have found that it serves as an excellent yardstick for determining potential mathematics majors.

Kenneth W. Anderson
Dick Wick Hall

ELEMENTARY
REAL ANALYSIS

Sets

1
Truth Tables and Operations with Sets

In the study of mathematics we start with certain primitive statements (called *axioms*) which we assume to be true throughout our entire discussion. We then use the elementary laws of logic to combine these statements to obtain new statements, known to be true, which we call *theorems*. Mathematicians have the enormous advantage of being able to define new concepts, study them on the basis of their intuitive notions, and then obtain theorems about these new concepts which may turn out to be important or enjoyable or both. There is always the exciting prospect of introducing a new primitive statement as an axiom to see what will happen when it is assumed to be true.

In order to give meaning to our theorems, we must talk about certain objects and collections of objects. Some of our terms must necessarily be undefined since we must start somewhere. We shall take the word "set" as one of our undefined terms, where a set means a collection of objects, and each member (or element) of the set is called a *point*. If A is a set, the notation $p \in A$ means that the point p is a member of the set A; the notation $p \notin A$ means that p is not a member of the set A. In order to define a set A we must be able to tell, given any point p, whether $p \in A$ or $p \notin A$. It is clear that for every point p, exactly one of the statements

(a) $$p \in A$$

(b) $$p \notin A$$

must be true.

The expression "for every point p" immediately introduces difficulty unless we have some type of universal set U consisting of all points that we wish to consider in a given discussion. Suppose we consider the following question: For how many values of x is

$$4x^4 + 55x^2 - 144 = 0$$

a true statement?

It is clear that in order to answer this question, x should be some kind of number. *What* kind is the important question. It is quite easy to rewrite our equation in the form

$$(4x^2 - 9)(x^2 + 16) = 0$$

If we take as our universe U, the set of all positive real numbers, then the equation has exactly one solution, $x = \frac{3}{2}$. If we take U as the set of all real numbers, there are two solutions, $x_1 = +\frac{3}{2}$ and $x_2 = -\frac{3}{2}$. It is an easy exercise to show that by taking a still different universe U, the equation will have four solutions. Note also that by taking U as the set of all real numbers x such that $x \geqslant 2$, the equation will have no solution.

We must thus agree that in any discussion we shall have a specific universe U in mind from which to extract the sets that interest us.

As an example, let us consider a specific universe $U = \{a,b,c\}$.

Table 1.1

Ways of assigning T and F to these elements	Element of U		
	a	b	c
First way	T	T	T
Second way	T	T	F
Third way	T	F	T
Fourth way	T	F	F
Fifth way	F	T	T
Sixth way	F	T	F
Seventh way	F	F	T
Eighth way	F	F	F

Suppose that we associate with each point of U one of the two values T and F. There are a number of ways in which this can be done, and we list them in Table 1.1.

We adopt a rather simple way of indicating which of the two letters T and F is assigned to each of the elements a, b, and c of U. Instead of the rather cumbersome statement "F is assigned to b when we assign letters the fourth way," we shall write the much shorter expression $\alpha_4(b) = F$. Using this terminology, Table 1.1 can be rewritten as follows:

$$
\begin{array}{lll}
\alpha_1(a) = T & \alpha_1(b) = T & \alpha_1(c) = T \\
\alpha_2(a) = T & \alpha_2(b) = T & \alpha_2(c) = F \\
\alpha_3(a) = T & \alpha_3(b) = F & \alpha_3(c) = T \\
\alpha_4(a) = T & \alpha_4(b) = F & \alpha_4(c) = F \\
\alpha_5(a) = F & \alpha_5(b) = T & \alpha_5(c) = T \\
\alpha_6(a) = F & \alpha_6(b) = T & \alpha_6(c) = F \\
\alpha_7(a) = F & \alpha_7(b) = F & \alpha_7(c) = T \\
\alpha_8(a) = F & \alpha_8(b) = F & \alpha_8(c) = F
\end{array}
$$

Suppose we think for a moment about the third way in which the values T and F are assigned to the elements of U. Note that we have $\alpha_3(a) = T$, $\alpha_3(b) = F$, and $\alpha_3(c) = T$. Although we shall give a somewhat more precise definition of a function later in the text, it is sufficient for our present purposes to reword our last remark by saying that α_3 is a function defined on U which assigns to a, b, c the respective values T, F, T. In exactly the same way we see that α_7 is a function defined on U which assigns to a, b, c the respective values F, F, T. Also, α_8 is a function defined on U which assigns to a, b, c the respective values F, F, F.

In general, for any universe U, a function which assigns exactly one of the two values T and F to each element of U is called a *truth function* defined on U. We refer to T and F as *truth values*.

Returning to the example above, let us consider the truth function α_5 defined on U. We see that $\alpha_5(a) = F$, $\alpha_5(b) = T$, $\alpha_5(c) = T$. The collection of all elements of U to which α_5 assigns the symbol T is clearly the set $\{b,c\}$. The set of all elements of U to which α_5 assigns the symbol F is the set $\{a\}$.

For any given universe U, a *subset* of U is any collection of elements of U. Note in particular that U may be considered as a subset of itself. In the above example, we see that the subset of U consisting of all elements of U to which α_5 assigns the value T is precisely the set $A = \{b,c\}$. The subset of U consisting of all elements of U to which α_5 assigns the value F is precisely the set $B = \{a\}$.

It is clear that if α is any truth function defined on a universe U,

Table 1.2

Truth-function	Subset of U determined by the truth function
α_1	U
α_2	$\{a,b\}$
α_3	$\{a,c\}$
α_4	$\{a\}$
α_5	$\{b,c\}$
α_6	$\{b\}$
α_7	$\{c\}$
α_8	$\varnothing$

then there is a subset A of U consisting of all elements of U to which α assigns the value T. We say that A is the subset of U determined by α.

Table 1.2 lists each of the eight truth functions defined on the universe $U = \{a,b,c\}$ together with the subset of U which each determines.

Notice the symbol opposite the truth function α_8. Since α_8 assigns the symbol F to every element of U, the subset of U determined by α_8 contains no elements. We call the set containing no elements the *empty set* and denote it by $\varnothing$.

In general, each truth function α defined on a universe U determines a unique, i.e., one and only one, subset A of U. For this reason it is convenient to denote this truth function by α_A. Thus, given a truth function α_A defined on U, α_A assigns the value T to each element of U in the set A, and α_A assigns the value F to each element of U which is not in the set A.

We have seen that each truth function α defined on U determines a unique subset A of U. On the other hand, given any subset A of U we can find exactly one truth function defined on U which determines the subset A of U. Clearly, this truth function α_A *must* be defined so that it assigns the value T to every element of U in the set A and the value F to every element of U not in the set A.

In view of the above discussion, it is clear that when we are discussing sets, we are free to use either the set itself or its truth function, since each is uniquely determined by the other.

It is important that we be able to combine truth functions to get other truth functions. In particular, if α and β are truth functions, then each of the expressions "α and β," "α or β," "α implies β" will be a truth function. To avoid the confusion which arises if we mix words with symbols, we adopt the following notation:

$$\alpha \text{ and } \beta \qquad \text{is denoted by} \qquad \alpha \wedge \beta$$

$$\alpha \text{ or } \beta \qquad \text{is denoted by} \qquad \alpha \vee \beta$$

$$\alpha \text{ implies } \beta \qquad \text{is denoted by} \qquad \alpha \Longrightarrow \beta$$

We have seen that a truth function α defined on a universe U satisfies either $\alpha(p) = T$ or $\alpha(p) = F$ for each point p of U. Thus if we know whether $\alpha(p) = T$ or $\alpha(p) = F$ and whether $\beta(p) = T$ or $\beta(p) = F$, we must be able to decide which of the two symbols T and F each of the truth functions $\alpha \wedge \beta$, $\alpha \vee \beta$, $\alpha \Rightarrow \beta$ assigns to each point p of our universe U. We do this by giving *truth tables*, which we regard as the *definitions* of the values of the expressions we are trying to determine. Note carefully that the truth table we give holds for each fixed p in the universe U.

A careful examination of Table 1.3 provides the following useful information:

$$(\alpha \wedge \beta)(p) = \begin{cases} T \\ F \end{cases} \quad \begin{array}{l} \text{when } both\ \alpha(p) = T \text{ and } \beta(p) = T \\ \text{otherwise} \end{array} \tag{1}$$

$$(\alpha \vee \beta)(p) = \begin{cases} F \\ T \end{cases} \quad \begin{array}{l} \text{when } both\ \alpha(p) = F \text{ and } \beta(p) = F \\ \text{otherwise} \end{array} \tag{2}$$

$$(\alpha \Longrightarrow \beta)(p) = \begin{cases} F \\ T \end{cases} \quad \begin{array}{l} \text{when } both\ \alpha(p) = T \text{ and } \beta(p) = F \\ \text{otherwise} \end{array} \tag{3}$$

Another way of expressing Eq. (1) is to say that $(\alpha \wedge \beta)(p)$ is assigned the truth value F whenever at least one of $\alpha(p)$ and $\beta(p)$ is assigned the truth value F.

Another way of expressing Eq. (2) is to say that $(\alpha \vee \beta)(p)$ is assigned the truth value T whenever at least one of $\alpha(p)$ and $\beta(p)$ is assigned the truth value T. Note that in everyday usage, the word "or" is ambiguous. In this text we shall always use it in the sense of "at least one" and never in the sense of "one and only one."

Another way of expressing Eq. (3) is to say that $(\alpha \Rightarrow \beta)(p)$ is assigned the truth value T whenever $\alpha(p) = F$ or $\beta(p) = T$.

In Table 1.4 we repeat part of Table 1.3 and define a new truth

Table 1.3

$\alpha(p)$	$\beta(p)$	$(\alpha \wedge \beta)(p)$	$(\alpha \vee \beta)(p)$	$(\alpha \Longrightarrow \beta)(p)$
T	T	T	T	T
T	F	F	T	F
F	T	F	T	T
F	F	F	F	T

Table 1.4

$\alpha(p)$	$\beta(p)$	$(\alpha \Longrightarrow \beta)(p)$	$(\beta \Longrightarrow \alpha)(p)$	$(\alpha \Longleftrightarrow \beta)(p)$
T	T	T	T	T
T	F	F	T	F
F	T	T	F	F
F	F	T	T	T

function $\alpha \Longleftrightarrow \beta$, which is read "$\alpha$ if and only if β" abbreviated "α iff β."

We note from Table 1.4 that $(\alpha \Longleftrightarrow \beta)(p) = $ T *when and only when* we have *both* $\alpha(p) = $ T *and* $\beta(p) = $ T or *both* $\alpha(p) = $ F *and* $\beta(p) = $ F; in other words, $(\alpha \Longleftrightarrow \beta)(p) = $ T *when and only when* $\alpha(p) = \beta(p)$. For this reason, we say that the truth functions α and β are *logically equivalent at the point p.*

In general, two truth functions are called *logically equivalent* provided they are logically equivalent at every point p of the universe U. Clearly, the truth functions γ and δ are logically equivalent provided $\gamma(p) = \delta(p)$ for every $p \in U$. We shall indicate that γ and δ are logically equivalent by writing $\gamma = \delta$. It should be evident that $\gamma = \delta$ provided that in the truth table constructed for these truth functions the columns under γ and δ are identical and that this is true for every point p in the universe U.

As a simple example we use a truth table to show that $(\alpha \Rightarrow \beta) \wedge (\beta \Rightarrow \alpha)$ is logically equivalent to $\alpha \Longleftrightarrow \beta$. We let p be any point of the universe and use part of Table 1.4 to construct the required truth table. Since the last two columns in Table 1.5 are identical, and since p is *any* point of U, the two expressions under consideration are logically equivalent.

As another example, we show in Table 1.6 that $\alpha \Rightarrow (\alpha \wedge \beta)$ is logically equivalent to $\alpha \Rightarrow \beta$ since the last two columns are identical.

We shall use the symbol Δ to represent the truth function defined on U by

$$\Delta(p) = \text{T} \qquad \text{for every } p \in U$$

Table 1.5

$\alpha(p)$	$\beta(p)$	$(\alpha \Longrightarrow \beta)(p)$	$(\beta \Longrightarrow \alpha)(p)$	$[(\alpha \Longrightarrow \beta) \wedge (\beta \Longrightarrow \alpha)](p)$	$(\alpha \Longleftrightarrow \beta)(p)$
T	T	T	T	T	T
T	F	F	T	F	F
F	T	T	F	F	F
F	F	T	T	T	T

Table 1.6

$\alpha(p)$	$\beta(p)$	$(\alpha \wedge \beta)(p)$	$[\alpha \Longrightarrow (\alpha \wedge \beta)](p)$	$(\alpha \Longrightarrow \beta)(p)$
T	T	T	T	T
T	F	F	F	F
F	T	F	T	T
F	F	F	T	T

As a final example, we use Table 1.7 to show that α and β are logically equivalent iff $\alpha \Longleftrightarrow \beta$ and Δ are logically equivalent. Symbolically, we want to prove that

$$(\alpha = \beta) \Longleftrightarrow [(\alpha \Longleftrightarrow \beta) = \Delta]$$

Mathematicians consider the following statements as equivalent; i.e., any two of them have precisely the same meaning. In each of these statements, S_1 and S_2 are truth functions. Any one of these statements may be written $S_1 \Rightarrow S_2$.

S_1 implies S_2.
S_1 is a sufficient condition for S_2.
S_2 is a necessary condition for S_1.
If S_1, then S_2.
S_2 if S_1.
S_1 only if S_2.

The typical definition in mathematics is a statement of the form "S_1 iff S_2." Such a statement has the following equivalent forms, any one of which may be written $S_1 \Longleftrightarrow S_2$:

S_1 iff S_2.
S_1 is a necessary and sufficient condition for S_2.
S_2 is a necessary and sufficient condition for S_1.
S_1 implies S_2, *and* S_2 implies S_1.

It will be useful to have a symbol for the *negation* of a truth func-

Table 1.7

$\alpha(p)$	$\beta(p)$	$(\alpha = \beta)(p)$	$(\alpha \Longleftrightarrow \beta)(p)$	$\Delta(p)$	$[(\alpha \Longleftrightarrow \beta) = \Delta](p)$
T	T	T	T	T	T
T	F	F	F	T	F
F	T	F	F	T	F
F	F	T	T	T	T

Table 1.8

$\alpha(p)$	$\alpha'(p)$
T	F
F	T

tion α. We use α' as this symbol, with the understanding that $\alpha'(p) = $ T whenever $\alpha(p) = $ F and $\alpha'(p) = $ F whenever $\alpha(p) = $ T. This definition is summarized in Table 1.8.

Proving several or all of the assertions in the following theorem will provide some worthwhile practice in working with truth tables.

Theorem   1.1

Let α, β, δ be truth functions defined on the universe U. Let q be any point of U. Then the following statements are true:

(a) If $\alpha(q)$ implies $\beta(q)$, and if $\beta(q)$ implies $\delta(q)$, then $\alpha(q)$ implies $\delta(q)$.
(b) $(\alpha \wedge \beta)(q)$ implies $\alpha(q)$.
(c) $(\alpha \wedge \beta)(q)$ implies $\beta(q)$.
(d) $\alpha(q)$ implies $(\alpha \vee \beta)(q)$.
(e) $\alpha \wedge \beta$ is logically equivalent to $\beta \wedge \alpha$.
(f) $\alpha \vee \beta$ is logically equivalent to $\beta \vee \alpha$.
(g) $\alpha \Rightarrow \beta$ is logically equivalent to $\beta' \Rightarrow \alpha'$.
(h) The truth function Δ' is defined on U by $\Delta'(p) = $ F for every $p \in U$.
(i) $\Delta'(q)$ implies $\alpha(q)$.

Using Theorem 1.1g, we see that for every point $p \in U$ the statements $(\alpha \Rightarrow \beta)(p)$ and $(\beta' \Rightarrow \alpha')(p)$ are logically equivalent. Each of these statements is called the *contrapositive* of the other. It is quite often the case in mathematics that a proof of a theorem can be obtained more easily by proving its contrapositive.

Theorem 1.2   De Morgan's theorem

Let α and β be truth functions defined on the universe U. Then:

(a) $(\alpha \wedge \beta)'$ is logically equivalent to $\alpha' \vee \beta'$.
(b) $(\alpha \vee \beta)'$ is logically equivalent to $\alpha' \wedge \beta'$.

Proof   We leave part (b) as an exercise (Prob. 1.3) and prove (a) by constructing a truth table. Let p be any point of U. Inasmuch as the fourth and seventh columns of Table 1.9 are identical, (a) is proved. $\square$

Table 1.9

$\alpha(p)$	$\beta(p)$	$(\alpha \wedge \beta)(p)$	$(\alpha \wedge \beta)'(p)$	$\alpha'(p)$	$\beta'(p)$	$(\alpha' \vee \beta')(p)$
T	T	T	F	F	F	F
T	F	F	T	F	T	T
F	T	F	T	T	F	T
F	F	F	T	F	T	T

We have seen earlier in this chapter that every subset of U corresponds to a truth function defined on the universe U and that every truth function defined on the universe U corresponds to a subset of U. In particular, a truth function α_A defined on U by $\alpha_A(p) = \text{T}$ iff $p \in A$ associates a unique subset A of U with the truth function α_A, and conversely. We can now use truth functions of this type to obtain a number of interesting properties of sets.

Definition 1.3

Let A and B be subsets of U. We say that $A = B$ iff $\alpha_A = \alpha_B$; that is, iff $(\alpha_A \iff \alpha_B) = \Delta$. The symbol $\neq$ will be used for the negation of equality.

Definition 1.4

Let A and B be sets in the universe U.

(a) The set $A \cup B$ is the unique set determined by the truth function $\alpha = \alpha_A \vee \alpha_B$. This set is called the *union* of the two sets A and B.
(b) The set $A \cap B$ is the unique set determined by the truth function $\beta = \alpha_A \wedge \alpha_B$. This set is called the *intersection* of the two sets A and B.
(c) The set $C(A)$, called the *complement* of the set A, is the unique set determined by the truth function α_A'.
(d) The set A is said to be contained in the set B iff $\alpha_A \Rightarrow \alpha_B$. To indicate that A is contained in B, we write $A \subset B$ (or, equivalently, $B \supset A$). We call A a *subset* of B and B a *superset* of A.
(e) A set A is a *proper* subset of B iff $A \subset B$, $A \neq \varnothing$, $A \neq B$.
(f) A *singleton* or *singleton set* is a set consisting of exactly one point p. We write the set as $\{p\}$.

Theorem 1.5

Let A, B, C be subsets of U. Then:

(a) If $A \subset B$ and $B \subset C$, then $A \subset C$.
(b) $\varnothing \subset A$.
(c) $A \cap B \subset A \subset A \cup B$.

(d) $A \cup B = B \cup A$.
(e) $A \cap B = B \cap A$.
(f) $A = B$ iff $A \subset B$ and $B \subset A$.
(g) A point p is in $A \cup B$ iff p is in at least one of these sets.
(h) A point p is in $A \cap B$ iff p is in both of these sets.
(i) $C(A \cup B) = C(A) \cap C(B)$.
(j) $C(A \cap B) = C(A) \cup C(B)$.

Proof   For (a) use Theorem 1.1a.
For (b) use Theorem 1.1i.
For (c) use Theorem 1.1b and 1.1d.
Parts (d) and (e) are immediate from the facts that

$$\alpha_A \vee \alpha_B = \alpha_B \vee \alpha_A \qquad \text{and} \qquad \alpha_A \wedge \alpha_B = \alpha_B \wedge \alpha_A$$

Part (f) is immediate from the fact, which we have proved, that

$$\alpha_A \Longleftrightarrow \alpha_B = [(\alpha_A \Longrightarrow \alpha_B) \wedge (\alpha_B \Longrightarrow \alpha_A)]$$

Parts (g) and (h) are left as exercises in Prob. 1.4.
Parts (i) and (j) are immediate from De Morgan's theorem.   □

Using Theorem 1.1g, our next theorem is immediate.

Theorem 1.6

Given any two sets A and B, $A \subset B$ iff $C(B) \subset C(A)$.

Theorem 1.7

For any sets A and B we have

(a) $A = A \cap B$ iff $A \subset B$
(b) $A = A \cup B$ iff $B \subset A$
(c) $A \cap B = A \cup B$ iff $A = B$

Proof   Part (c) is immediate from (a) and (b). The proof of (b) is an exercise. Although (a) is immediate from Definition 1.3 and Table 1.7, we shall give another proof utilizing De Morgan's theorem. From Theorem 1.5c, $A \cap B \subset A$. Thus the proof of (a) will be complete if we show that $A \subset A \cap B$ iff $A \subset B$. Using Theorems 1.6 and 1.5j, we see that

$$A \subset A \cap B \quad \text{iff } C(A \cap B) \subset C(A)$$
$$\text{iff } C(A) \cup C(B) \subset C(A)$$
$$\text{iff } C(B) \subset C(A)$$
$$\text{iff } A \subset B \quad □$$

Using the fact that every truth function is the truth function of a set, we have the following corollary.

Corollary 1.8

Let α and β be truth functions defined on U. Then

(a) $\alpha \implies \alpha \wedge \beta$ iff $\alpha \implies \beta$

(b) $\alpha \iff \alpha \vee \beta$ iff $\beta \implies \alpha$

(c) $(\alpha \iff \beta) \iff [(\alpha \wedge \beta) \iff (\alpha \vee \beta)]$

Definition 1.9

For any sets A and B in a universe U, we denote by $A - B$ the set of all points of U which belong to A but do not belong to B; that is,

$$A - B = A \cap C(B)$$

We sometimes refer to $A - B$ as the *complement of the set B with respect to the set A.*

If the set A of Definition 1.9 is chosen as the universe U, we have $U - B = U \cap C(B) = C(B)$, so that $U - B$ is an alternate form of expressing the complement of the set B with respect to the universe U.

We shall accept the word "collection" as a synonym for "set" in our mathematical language, because it seems more reasonable to speak of a "collection of sets" than a "set of sets."

Definition 1.10

(a) The sets A and B are said to be *disjoint* iff their intersection is the empty set.

(b) A collection of sets is said to be *disjoint* iff no two sets of the collection have a point in common.

We conclude this chapter with an additional set notation which is extremely useful, particularly as a device for defining sets. As we have already seen a number of times, a set is generally characterized in terms of some distinguishing property or properties shared by all its members. Suppose we denote such a property (or combination of properties) by P and for each point $x \in U$ write $P(x)$ for the sentence "x satisfies P." Clearly, this makes P a truth function, and this truth function determines a set. We can thus *define* a set in this way, and the notation $\{x \in U \mid P(x)\}$ means "the set of all points x in the universe U such that $P(x)$ is true." When the universe is understood, we sometimes shorten the notation to $\{x \mid P(x)\}$.

This new notation can now be used to summarize the various

types of sets which we have introduced:

$$A \cap B = \{x \in U \mid x \in A \text{ and } x \in B\}$$
$$A \cup B = \{x \in U \mid x \in A \text{ or } x \in B\}$$
$$C(A) = \{x \in U \mid x \notin A\}$$
$$A - B = \{x \in U \mid x \in A \text{ and } x \notin B\}$$

PROBLEMS

1.1 Describe a universe U in which the equation

$$(4x^2 - 9)(x^2 + 16) = 0$$

has four solutions.

1.2 Use truth tables to prove several parts of Theorem 1.1.

1.3 Prove Theorem 1.2b (a) using truth tables and (b) without using truth tables.

1.4 Fill in all missing details in the proof of Theorem 1.5. In particular give proofs of (g) and (h) of this theorem.

1.5 Show how Corollary 1.8 follows from Theorem 1.7. Then give a proof of this corollary using truth tables. Finally, show that if we have proved Corollary 1.8, we can obtain Theorem 1.7 as a corollary.

1.6 Show that for any set A, $C[C(A)] = A$.

1.7 Let $U = \{a,b,c,d\}$.
    (a) Find all subsets of U. How many are there?
    (b) How many proper subsets are there of U?
    (c) If we enlarge U by adding another point, what is the effect on the total number of subsets?
    (d) How many subsets of U are singletons?

1.8 In this problem we have a universe $U = \{a,b,c,d,e,f,g\}$ and subsets $A = \{a,b,e\}$, $B = \{b,c,d\}$, $C = \{c,e,g\}$, $D = \{a,b,d,f\}$.
    (a) Find the following:

(i) $A \cup B$	(ii) $C \cup D$
(iii) $A \cap B$	(iv) $C \cap D$
(v) $C(A)$	(vi) $C(B)$
(vii) $B - A$	(viii) $A - B$
(ix) $(A - B) \cup (B - A)$	(x) $(A \cup B) - (A \cap B)$

    (b) What can be said about the sets C and D?
    (c) Show that the sets $(A \cap B) \cup C$ and $A \cap (B \cup C)$ are not equal.

1.9 Let A, B, C, H be sets in a universe U. Show that the following statements are true:
    (a) Distributive law for intersections over unions:

$$A \cap (B \cup C) = (A \cap B) \cup (A \cap C)$$

    (b) Distributive law for unions over intersections:

$$A \cup (B \cap C) = (A \cup B) \cap (A \cup C)$$

(c) If $A \subset B$, then $A \cap H \subset B \cap H$ and $A \cup H \subset B \cup H$.

(d) $\qquad\qquad C - (A \cup B) = (C - A) \cap (C - B)$

(e) $\qquad\qquad C - (A \cap B) = (C - A) \cup (C - B)$

1.10   State in words the various assertions of Theorem 1.5.

1.11   The equations in Probs. 1.9d and 1.9e extend De Morgan's laws to three sets. Show that De Morgan's laws are the special case of these equations in which the set C is chosen as the universe U.

1.12   Show that for any sets A and B,

$$(A - B) \cup (B - A) = (A \cup B) - (A \cap B)$$

The set denoted by either of these equal expressions is called the *symmetric difference* of the sets A and B.

1.13   What can we conclude about two sets whose symmetric difference is the empty set?

1.14   What can we conclude about the set B if the symmetric difference of A and B is the set A?

1.15   State carefully in words the result of Prob. 1.12.

1.16   State precisely (using intersections) what is meant by the assertion that a collection $\{A,B,C\}$ of three sets is disjoint.   Show that if a collection $\{A,B,C\}$ is disjoint, then $A \cap B \cap C = \varnothing$.   Give an example of three sets A, B, C which satisfy $A \cap B \cap C = \varnothing$ but such that the collection $\{A,B,C\}$ is not disjoint.

1.17   Devise a reasonable definition of the union and of the intersection of 257 sets A_1, $A_2, A_3, \ldots, A_{257}$.   Guess at a generalization of De Morgan's laws, and try to prove that you are correct.   Do not use mathematical induction.

2

The Real Number System and Mathematical Induction

Since our universe throughout most of this text will be the set R_1 of real numbers, this chapter is devoted to a brief discussion of the real number system. It is possible to carry out a careful "construction" of the real number system, beginning with a minimal set of axioms for the set I of positive integers. This process of construction is both interesting and informative, but it is so lengthy that it almost constitutes a course in itself. The interested reader will find such a development of the real number system in most texts on modern algebra. We choose instead the *axiomatic approach*, wherein the real number system is defined in terms of certain basic properties which it satisfies, and we shall accept these basic properties as axioms.

Modern algebra provides the terminology for an elegant description of the real number system in just three words: a *complete ordered field*. Of course, each of these words requires definition before this description conveys meaning. We choose to bypass the formal abstract definitions of these terms and instead list their properties only as they apply directly to R_1. We do so in two stages, introducing the ordered field axiom here and accounting for the word "complete" in Chap. 4, with the introduction of the least upper bound axiom.

Ordered field axiom

The real number system is an ordered field.

We shall now state explicitly the assumptions which are implicit in the ordered field axiom, beginning with the term "field."

This axiom asserts that R_1 is a *field* under the operations of addition and multiplication. Specifically, for any $x, y \in R_1$, the sum $x + y$ and the product xy are uniquely determined elements of R_1, and the following conditions are satisfied:

(i) $\left. \begin{array}{l} x + y = y + x \\ \quad \text{and} \\ xy = yx \end{array} \right\}$ for all $x, y \in R_1$ commutative laws

(ii) $\left. \begin{array}{l} x + (y + z) = (x + y) + z \\ \quad \text{and} \\ x(yz) = (xy)z \end{array} \right\}$ for all $x, y, z \in R_1$ associative laws

(iii) $x(y + z) = xy + xz$ for $x, y, z \in R_1$ distributive law

(iv) There exist distinct elements 0 and 1 in R_1 such that
$\left. \begin{array}{l} x + 0 = x \\ \quad \text{and} \\ x \cdot 1 = x \end{array} \right\}$ for all $x \in R_1$ identity elements

(v) For each $x \in R_1$ there exists $y \in R_1$ such that
$x + y = 0$ additive inverse

(vi) For each $x \in R_1$ such that $x \neq 0$,
there exists $y \in R_1$ such that $xy = 1$ multiplicative inverse

The reader is undoubtedly familiar with the commutative and associative laws, but the remaining properties deserve some comment. Another distributive law follows from (iii) and the commutativity of multiplication, since

$$(x + y)z = z(x + y) = zx + zy = xz + yz$$

The element 0 whose existence is asserted in (iv) also satisfies

$$0 + x = x$$

by virtue of the commutativity of addition. It is called the *additive identity* and is easily seen to be unique since, if $0'$ were another element of R_1 satisfying

$$x + 0' = 0' + x = x \qquad \text{for every } x \in R_1 \tag{1}$$

then substitution of $0'$ for x in (iv) and 0 for x in (1) would yield

$$0' = 0' + 0 = 0$$

The element 1 whose existence is asserted in (iv) also satisfies

$$1 \cdot x = x$$

and is called the *multiplicative identity*. We leave the proof of its uniqueness as an exercise.

Condition (v) asserts that every real number has an additive inverse, and its uniqueness is easily established. For suppose y and z are real numbers satisfying $x + y = 0$ and $x + z = 0$. Using (iv) and (ii), we have

$$y = y + 0 = y + (x + z) = (y + x) + z = 0 + z = z$$

We normally denote the unique additive inverse of a real number x by $-x$ and simplify our notation somewhat by defining *subtraction:*

$$x - y = x + (-y) \qquad \text{for } x, y \in R_1$$

Note that 0 is its own additive inverse since $0 + 0 = 0$. We shall show below in Theorem 2.1 that 0 is the only real number with this property.

Condition (vi) asserts that every nonzero real number has a multiplicative inverse, and we leave the proof of its uniqueness as an exercise. We normally denote the unique multiplicative inverse of a nonzero real number x by $1/x$ and define *division* as a notational convenience:

$$\frac{x}{y} = x \cdot \frac{1}{y} \qquad \text{for } x, y \in R_1, y \neq 0$$

It is easy to see that 1 is its own multiplicative inverse, since $1 \cdot 1 = 1$. We shall show below that $(-1)(-1) = 1$ and hence that -1 is also its own multiplicative inverse; furthermore, 1 and -1 are the only real numbers with this property.

Now that we know the field properties of R_1, we concern ourselves with the term "order." The ordered field axiom asserts that the field R_1 of real numbers is an *ordered* field. Specifically, this means that R_1 contains a set P of *positive elements* satisfying the following conditions:

(vii)   If $x, y \in P$, then $x + y \in P$ and $xy \in P$.
(viii)   For each $x \in R_1$ exactly one of the following is true:

$$x \in P \qquad -x \in P \qquad x = 0$$

We may now use the set P of positive elements to define an order in R_1. For $x, y \in R_1$, we shall say that x is *less than* y and write $x < y$ (or equivalently, that y is *greater than* x and write $y > x$) iff $y - x \in P$.

Note that for every $x \in P$ we have $x - 0 \in P$ and hence $x > 0$. Thus we may write $P = \{x \in R_1 \mid x > 0\}$. Similarly, if $-x \in P$, then $0 - x \in P$, so that $x < 0$. We call P the set of *positive real numbers* and the set $-P = \{x \in R_1 \mid -x \in P\}$ the set of *negative real numbers*. It is clear from (viii) above, which is called the *law of trichotomy*, that R_1 is the union of the three disjoint sets $P, -P, \{0\}$. Furthermore, for each pair of real numbers x, y, exactly one of the following is true: $x > y$, $x < y$, $x = y$, since by (viii), $x - y \in P$, $y - x \in P$, or $x - y = 0$.

It is convenient to introduce the notation $x \geqslant y$ (or equivalently, $y \leqslant x$) to mean "$x > y$ or $x = y$." Clearly, if $x \geqslant y$ and $y \geqslant x$, then $x = y$.

This concludes our brief explanation of the basic conditions on R_1 asserted by the ordered field axiom. Since it is not our intention to undertake an exhaustive study of the algebraic and order properties of R_1, we merely cite a few interesting theorems to illustrate some techniques of proof using the axiom.

Our first theorem states that 0 is the only real number which is its own additive inverse.

Theorem 2.1

If $x \in R_1$ and $x = -x$, then $x = 0$.

Proof   The contrapositive is an immediate consequence of (viii), since if $x \neq 0$, we must have either $x \in P$ or $-x \in P$, but not both, and hence $x \neq -x$. $\square$

Theorem 2.2   Cancellation law for addition

If $x, y, z \in R_1$, and if $x + z = y + z$, then $x = y$.

Proof   Using (iv), (ii), and the fact that $z + (-z) = 0$, we have

$$x = x + 0 = x + [z + (-z)] = (x + z) + (-z)$$
$$= (y + z) + (-z) = y + [z + (-z)] = y + 0 = y \qquad \square$$

Corollary 2.3

For each $x \in R_1$, $-(-x) = x$.

Proof   Since $x + (-x) = 0$ and $-x + [-(-x)] = 0$, we have $-x + x = -x + [-(-x)]$, and the result follows from Theorem 2.2. $\square$

Theorem 2.4

For every $x \in R_1$, $x \cdot 0 = 0$.

Proof   Using the distributive law, we have

$$x + x \cdot 0 = x \cdot 1 + x \cdot 0 = x(1 + 0) = x \cdot 1 = x = x + 0$$

and the result follows from Theorem 2.2   □

Theorem 2.5

For $x, y \in R_1$, $xy = 0$ iff $x = 0$ or $y = 0$.

Proof   If $x = 0$ or $y = 0$, then $xy = 0$ by Theorem 2.4.   For the converse, suppose $xy = 0$ and $x \neq 0$.   Then we have

$$y = 1 \cdot y = \left(\frac{1}{x} \cdot x\right) y = \frac{1}{x}(xy) = \frac{1}{x} \cdot 0 = 0   □$$

Theorem 2.6

For $x \in R_1$, $(-1)x = -x$.

Proof   The reader may verify that for each $x \in R_1$,

$$x + (-x) = 0 = 0 \cdot x = [1 + (-1)] \cdot x = 1 \cdot x + (-1)x = x + (-1)x$$

and the result follows from Theorem 2.2.   □

Corollary 2.7

$$(-1)(-1) = 1$$

Proof   Choosing $x = -1$ in Theorem 2.6, we have $(-1)(-1) = -(-1)$, and the result follows from Corollary 2.3.   □

Corollary 2.7 asserts that -1 is its own multiplicative inverse. We now show that 1 and -1 are the only real numbers with this property.

Theorem 2.8

If $x \in R_1$ and $x = 1/x$, then $x = 1$ or $x = -1$.

Proof   An easy application of the distributive laws verifies that

$$(x + 1)(x - 1) = x \cdot x - 1$$

Since $x = 1/x$, we have $x \cdot x = x \cdot 1/x = 1$, and hence $x \cdot x - 1 = 0$. Thus $(x + 1)(x - 1) = 0$, and by Theorem 2.5, either $x + 1 = 0$, in which case $x = -1$, or $x - 1 = 0$, in which case $x = 1$.   □

Theorem 2.9

For $x,y,z \in R_1$, if $x < y$ and $y < z$, then $x < z$.

Proof Note that $z - x = (z - y) + (y - x) \in P$ by (vii), since $z - y \in P$ and $y - x \in P$ by hypothesis   □

Theorem 2.10

For $x,y,z \in R_1$, if $x < y$, then $x + z < y + z$.

Proof Immediate, since

$$(y + z) - (x + z) = y + z - x - z = y - x > 0   □$$

Theorem 2.11

Let $x,y,z \in R_1$. Then:

(a) If $x < y$ and $z > 0$, then $xz < yz$.
(b) If $x < y$ and $z < 0$, then $xz > yz$.

Proof Since $yz - xz = (y - x)z$, (a) follows immediately from (vii). For (b), note that $y - x \in P$ and $-z \in P$, and hence by (vii) and the distributive law

$$xz - yz = (y - x)(-z) \in P   □$$

Additional algebraic and order properties of the real numbers are included in the problems. We now turn our attention to certain subsets of R_1 which also play important roles in analysis, beginning with the set I of positive integers (or natural numbers). Although we are fairly familiar with most of the properties of the positive integers, it seems appropriate to list our basic assumptions about this important set in an axiom.

Positive integers axiom

The set I of positive integers is a subset of the set P of positive real numbers satisfying the following conditions:

(a) $1 \in I$.
(b) If $m,n \in I$, then $m + n \in I$ and $mn \in I$.
(c) If $k \in I$ and $k \neq 1$, then $k - 1 \in I$.
(d) *Well-ordering principle* Every nonempty subset of I contains a smallest element.

Since I is a subset of R_1, the commutative, associative, and distributive laws hold in I, and by virtue of (a), the positive integer 1 serves as the multiplicative identity for I as well as for R_1. However, I has no additive identity, since $0 \notin I$; nor do (v) or (vi) of the ordered field axiom hold in I. According to (b) above, the sum and product of two positive integers are positive integers. As a special case of (b), note that if $n \in I$, then $n + 1 \in I$. We call $n + 1$ the *successor* of the positive integer n, and it is evident that every positive integer has a successor. It should be equally clear that 1 is the only positive integer which is not a successor of a positive integer, for if $x + 1 = 1 = 0 + 1$, it follows from Theorem 2.2 that $x = 0$. But zero is not a positive integer.

By adjoining to the set I the set $-I = \{n \mid -n \in I\}$ of negative integers and the singleton $\{0\}$, we get the set Z of all integers. It can be shown that the sum and product of two elements of Z are again elements of Z, and that conditions (i) through (v) of the ordered field axiom hold in Z. However, Z is not a field since it fails to satisfy (vi). The set Z also inherits a *natural ordering* from the ordering of R_1,

$$\cdots -3 < -2 < -1 < 0 < 1 < 2 < 3 < \cdots$$

and we see that if $m,n \in Z$, then $m > n$ iff $m - n \in I$.

Our next theorem establishes the principle of mathematical induction, an extremely important technique which will be used extensively throughout this text.

Theorem 2.12   The principle of mathematical induction

Let S be any subset of I satisfying the following conditions:

(a) $1 \in S$.
(b)   If k is any positive integer such that $k \in S$, then $k + 1 \in S$.

Then $S = I$.

Proof   We prove the contrapositive by showing that if S is a subset of I such that $S \neq I$, then (a) and (b) cannot both be true. Since $S \subset I$ and $S \neq I$, the set $K = I - S$ is nonempty, and hence there is a smallest element $k \in K$, by the well-ordering principle. If $k = 1$, then $1 \notin S$, and so (a) fails. If $k \neq 1$, then $k - 1 \in I - K = S$, whereas $(k - 1) + 1 = k \notin S$, and so (b) fails. $\square$

A careful review of the proof of Theorem 2.12 should convince us that all four conditions (a), (b), (c) and (d) of the positive integers axiom are used in the proof. Thus, given (a), (b) and (c) as hypotheses, we may assert that the well-ordering principle implies the principle of mathematical induction. In the next chapter we shall prove the converse—that given (a), (b), and (c) as hypotheses, the principle of mathematical induction implies the well-ordering principle—and hence establish the equivalence of these two important principles. By virtue of this equivalence, either principle could be stated in (d).

In the remainder of this chapter we illustrate some useful techniques involving mathematical induction and at the same time derive results which will be useful in our later work. We classify the two primary uses of this principle as *inductive definitions* and *inductive proofs* and begin with an example of the former.

Example 2.13

For any real number x, we define $x^0 = 1$, $x^1 = x$, and for each integer $n \geq 1$, $x^{n+1} = x^n \cdot x$. We assert that x^n is now defined for any real number x and all nonnegative integers n. Clearly, x^n is defined for $n = 0$, and if we let S be the set of all positive integers n for which x^n is defined, it is easy to see that (a) and (b) of Theorem 2.12 are both satisfied, so that $S = I$.

Theorem 2.14

If x is a real number and n is a positive integer, then $x^n = 0$ iff $x = 0$.

Proof Let S be the set of all positive integers n for which $x^n = 0$ iff $x = 0$. Since $x^1 = x$, it is clear that $1 \in S$. Now suppose that $k \in S$. Since $x^{k+1} = x^k \cdot x$, it follows from Theorem 2.5 that $x^{k+1} = 0$ iff $x^k = 0$ or $x = 0$. But the fact that $k \in S$ implies that $x^k = 0$ iff $x = 0$. Thus, $x^{k+1} = 0$ iff $x = 0$, so that $k + 1 \in S$. Therefore $S = I$ by Theorem 2.12. $\square$

By virtue of Theorem 2.14, if $x \neq 0$, then $x^n \neq 0$ for any $n \in I$, and we may thus define

$$x^{-n} = \frac{1}{x^n} \qquad \text{for each } n \in I, \ x \neq 0$$

Now x^n is defined for any nonzero real number x and every $n \in Z$.

Theorem 2.15

If a and r are real numbers with $r \neq 1$, then for any positive integer n,

$$a + ar + ar^2 + ar^3 + \cdots + ar^{n-1} = \frac{a(1 - r^n)}{1 - r} \qquad (1)$$

Proof Define $s_1 = a$, and for each $n > 1$ define $s_n = s_{n-1} + ar^{n-1}$. We are to prove that

$$s_n = \frac{a(1 - r^n)}{1 - r} \qquad \text{for each } n \in I \tag{2}$$

Let S be the set of all positive integers n for which (2) is true. Clearly $1 \in S$, since $s_1 = a$. Now suppose that the positive integer $k \in S$. Then we have

$$s_k = \frac{a(1 - r^k)}{1 - r}$$

and hence it follows that

$$s_{k+1} = s_k + ar^k = \frac{a(1 - r^k)}{1 - r} + ar^k$$

$$= \frac{a - ar^k + ar^k - ar^{k+1}}{1 - r} = \frac{a(1 - r^{k+1})}{1 - r}$$

Thus $k + 1 \in S$. Therefore $S = I$ by Theorem 2.12. $\square$

The student may recall from his work in algebra that the expression on the left side of Eq. (1) is called a *geometric progression* with first term a and common ratio r. We have proved in Theorem 2.15 that the expression on the right side of Eq. (1) is the formula for the sum of the first n terms of such a geometric progression, where n is any positive integer.

We close this chapter by listing several interesting corollaries to Theorem 2.15.

Corollary 2.16

For any real number r and any positive integer n,

$$1 - r^n = (1 - r)(1 + r + r^2 + \cdots + r^{n-1})$$

Proof The result is immediate if $r = 1$, since both sides are zero (see Prob. 2.10). If $r \neq 1$, choose $a = 1$ in Theorem 2.15 and multiply both sides of (1) by $1 - r$. $\square$

Corollary 2.17

For any real numbers a and b and any positive integer n,

$$a^n - b^n = (a - b)(a^{n-1} + a^{n-2}b + a^{n-3}b^2 + \cdots + ab^{n-2} + b^{n-1})$$

Proof The result is immediate if $a = 0$. If $a \neq 0$, choose $r = b/a$ in Corollary 2.16 and multiply both sides by a^n. $\square$

Corollary 2.18

Let a and r be real numbers such that $a > 0$ and $0 < r < 1$. If n is any positive integer, then

$$a + ar + ar^2 + \cdots + ar^{n-1} < \frac{a}{1-r}$$

Proof Since $0 < r < 1$, Prob. 2.11 asserts that $0 < r^n < 1$, and hence we have $0 < 1 - r^n < 1$. Also, $a/(1-r) > 0$, and so by Theorem 2.11a,

$$(1 - r^n) \cdot \frac{a}{1-r} < 1 \cdot \frac{a}{1-r} = \frac{a}{1-r}$$

The result then follows from (1) of Theorem 2.15.   □

Corollary 2.19

Let ϵ be a real number such that $0 < \epsilon < 1$, and let n be any positive integer. Then

$$\frac{\epsilon}{2} + \left(\frac{\epsilon}{2}\right)^2 + \left(\frac{\epsilon}{2}\right)^3 + \cdots + \left(\frac{\epsilon}{2}\right)^n < \epsilon$$

Proof Choose $r = \epsilon/2$ and $a = \epsilon/2$ in Corollary 2.18, noting that

$$\frac{a}{1-r} = \frac{\epsilon/2}{1 - \epsilon/2} = \frac{\epsilon}{2 - \epsilon} < \epsilon \qquad □$$

PROBLEMS

2.1 Prove that the multiplicative identity asserted in (iv) is unique.

2.2 Prove that the multiplicative inverse of a nonzero real number is unique.

2.3 Prove the *cancellation law* for *multiplication:* If $x,y,z \in R_1$ with $z \neq 0$, and if $xz = yz$, then $x = y$.

2.4 Use the distributive laws to prove that $(x + 1)(x - 1) = x \cdot x - 1$.

2.5 Prove that if $x \in R_1$ and $x \neq 0$, then $x^2 > 0$.

2.6 Prove:
  (*a*) $1 > 0$. *Hint:* Use Prob. 2.5.
  (*b*) Use part (*c*) of the positive integers axiom to prove that 1 is the smallest positive integer.
  (*c*) Show that there is no positive integer between 1 and 2. Then show that if n is any positive integer, there is no positive integer between n and $n + 1$.
  (*d*) Conclude that if K is a nonempty subset of I and if $m \in I$ satisfies

(i) $m \leq x$ for every $x \in K$

and

(ii) $m \notin K$

then

$$m + 1 \leq x \qquad \text{for every } x \in K$$

2.7 Prove that if $x > 0$, then $1/x > 0$.

2.8 Prove that if $x \leqslant y$ and $y < z$, then $x < z$; also, if $x < y$ and $y \leqslant z$, then $x < z$.

2.9 Prove that if $0 < x < 1$, then $0 < x/2 < x < (x+1)/2 < 1$.

2.10 Use mathematical induction to prove that for every positive integer n, $1^n = 1$.

2.11 Use mathematical induction to prove that if $0 < x < 1$, then for every positive integer n, $0 < x^n \leqslant x$, with equality holding only for $n = 1$.

2.12 Prove that if $0 < x < 1$, then $x/(2 - x) < x$.

2.13 Show that a universe U containing n elements, where n is a positive integer, has 2^n subsets.

3
Bounded Sets and Absolute Value

From this point on our universe will be R_1 unless explicitly stated otherwise. Since R_1 is an ordered field, we may use its ordering to define bounded and unbounded sets of real numbers.

Definition 3.1

A real number b is said to be an *upper bound* for a set A in R_1 iff $x \leqslant b$ for every x in A, and a real number c is said to be a *lower bound* for A iff $x \geqslant c$ for every x in A. A set in R_1 is *bounded above* iff it has an upper bound, *bounded below* iff it has a lower bound, and *bounded* iff it is bounded both above and below. A set which is not bounded is called *unbounded*.

Example 3.2

Let $A = \{1,2,3\}$. It is easy to see that 1 is a lower bound for A and 3 is an upper bound for A; hence A is a bounded set. It follows from Prob. 2.8 that any real number less than 1 is also a lower bound for A and any real number greater than 3 is also an upper bound for A. However, 1 is the only lower bound for A which belongs to A, and 3 is the only upper bound for A which belongs to A.

In the technique of contrapositive proof we are interested in the negation of properties. For instance, we might ask: Exactly what is meant by the assertion that a real number is not an upper (or a lower) bound for a set? The general process of negating definitions is subtle, and sometimes difficult, but the present case is relatively simple, and we state it as a theorem.

Theorem 3.3

A real number b is not an upper bound for the set A in R_1 iff there exists an element x in A such that $x > b$, and a real number c is not a lower bound for A iff there exists an element x in A such that $x < c$.

Since the empty set $\varnothing$ contains no elements, our next result is an immediate consequence of Theorem 3.3.

Corollary 3.4

The empty set $\varnothing$ is bounded; in fact, every real number is both an upper bound and a lower bound for $\varnothing$.

We now give an example of a nonempty bounded set which contains none of its upper bounds or lower bounds.

Example 3.5

Let $A = \{x \in R_1 \mid 0 < x < 1\}$. Then 1 (or any real number greater than 1) is an upper bound for A, and 0 (or any real number less than 0) is a lower bound for A. We shall show that A contains none of its upper or lower bounds by proving the contrapositive: No member of the set A is a lower bound or an upper bound for A. Let b be any element in A. Since $0 < b < 1$, it follows from Prob. 2.9 that

$$0 < \frac{b}{2} < b < \frac{b+1}{2} < 1$$

and the result is an immediate consequence of Theorem 3.3.

Our next theorem establishes the equivalence of the well-ordering principle and the principle of mathematical induction (see Theorem 2.12 and the subsequent discussion).

Theorem 3.6

Given conditions (a), (b), and (c) of the positive integers axiom, the principle of mathematical induction implies the well-ordering principle.

Proof Let K be any subset of I. We want to show that if K is nonempty, then K has a smallest element, and we shall prove the contrapositive. Thus suppose K has no smallest element, and define the subset S of I as follows:

$$S = \{n \in I - K \mid n \text{ is a lower bound for } K\}$$

By Prob. 2.6, $1 \in S$. Now suppose k is an element of S. Since k is a lower bound for K which does not belong to K, it follows from Prob. 2.6 that $k + 1$ is a lower bound for K. Then $k + 1 \notin K$, since K has no smallest element, and hence $k + 1 \in S$. By the principle of mathematical induction, we thus have $S = I$, and hence $K = \varnothing$. □

The remainder of this chapter is devoted to a discussion of the absolute value of a real number. Note that our definition encompasses the entire set of real numbers by virtue of the law of trichotomy.

Definition 3.7

For $x \in R_1$, the *absolute value* of x, written $|x|$, is defined as follows:

$$|x| = \begin{cases} x & \text{if } x \geqslant 0 \\ -x & \text{if } x < 0 \end{cases}$$

Thus, for example, since 2 is positive, $|2| = 2$; whereas since -2 is negative, $|-2| = -(-2) = 2$. The absolute value of a real number is therefore its numerical value, neglecting sign. If we interpret the real numbers as points on the real line, absolute value has a geometrical significance; that is, $|x|$ merely denotes the *undirected distance* between the origin and the point x. Consider the following example.

Example 3.8

Let $A = \{x \mid |x| < 2\}$. Now, in terms of the geometrical significance of absolute value, we see that the set A consists of all those points whose undirected distance from the origin is less than 2. From this fact, we conclude that $A = \{x \mid -2 < x < 2\}$. Let us verify this result analytically, using the definition of absolute value. If $x \geqslant 0$, then $|x| = x$, and so the condition on A for nonnegative x becomes $x < 2$. If $x < 0$, then $|x| = -x$, and so the condition on A for negative x becomes $-x < 2$. By Theorem 2.11b, this last inequality (after multiplying through by -1) becomes $x > -2$. Combining the two results, we thus have $-2 < x < 2$ if $x \in A$; that is,

$$A = \{x \mid -2 < x < 2\}$$

Note that the same basic technique would yield a corresponding result if we were to replace 2 in Example 3.8 by any other positive number, and also if we were to allow for equality instead of the strict inequality. We thus have the following result, which we express as a lemma.

Lemma 3.9

For any $c > 0$,

$$\{x \mid |x| < c\} = \{x \mid -c < x < c\}$$

and for any $c \geq 0$,

$$\{x \mid |x| \leq c\} = \{x \mid -c \leq x \leq c\}$$

Just as $|x|$ denotes the undirected distance between the origin and the point x, so $|x - \alpha|$ denotes the undirected distance between the point x and the point α.

Example 3.10

Let $A = \{x \mid |x - 3| < 2\}$. Here, we see that A consists of all those points whose undirected distance from the point 3 is less than 2. From this fact, we conclude that $A = \{x \mid 1 < x < 5\}$. We can again verify this analytically, but instead of definition of absolute value we shall use Lemma 3.9, replacing x by $x - 3$ and c by 2. This yields $-2 < x - 3 < 2$. Adding 3 to each term of these inequalities, we obtain

$$-2 + 3 < x - 3 + 3 < 2 + 3 \qquad \text{or} \qquad 1 < x < 5$$

the desired result.

Note that the same basic technique would apply in Example 3.10 if 3 were replaced by any real number and 2 were replaced by any positive number, as well as if equality were allowed rather than strict inequality.

Lemma 3.11

Let $a,b \in R_1$. Then, if $b > 0$,

$$\{x \mid |x - a| < b\} = \{x \mid a - b < x < a + b\}$$

and if $b \geqslant 0$,

$$\{x \mid |x - a| \leqslant b\} = \{x \mid a - b \leqslant x \leqslant a + b\}$$

Having arrived at this point, we can look back and see that Lemma 3.9 is merely the special case of Lemma 3.11 where $a = 0$. It should also be evident that any set whose points satisfy an inequality of the form of those in Lemma 3.11 is necessarily bounded. In fact, if $A = \{x \mid |x - a| \leqslant b\}$, where $a,b \in R_1$ and $b \geqslant 0$, Lemma 3.11 tells us that $a + b$ is an upper bound for A and $a - b$ is a lower bound for A.

We now establish some of the important properties of absolute value, which we combine into a single theorem. Some of the proofs are carried out in complete detail to serve as a guide, others are sketched, and still others are left as exercises.

Theorem 3.12

Let $a,b \in R_1$. Then the following statements hold:

(a) $$|-a| = |a|$$

(b) $$-|a| \leqslant a \leqslant |a|$$

(c) $$|ab| = |a| \cdot |b|$$

(d) $$|a|^2 = a^2$$

(e) $$ab \leqslant |a| \cdot |b|$$

(f) $$\left|\frac{a}{b}\right| = \frac{|a|}{|b|} \text{ provided } b \neq 0$$

(g) $$|a + b| \leqslant |a| + |b| \qquad \text{triangle inequality}$$

(h) $$|a - b| \leqslant |a| + |b|$$

(i) $$|a - b| \geqslant ||a| - |b||$$

Proof All nine statements follow easily if $a = 0$. We thus assume $a \neq 0$. Similarly, we may assume $b \neq 0$.

(a) If $a > 0$, then $-a < 0$; therefore $|a| = a$, and $|-a| = -(-a) = a$. If $a < 0$, then $-a > 0$; therefore $|a| = -a$, and $|-a| = -a$. Thus, for every choice of a, $|-a| = |a|$.

(b) If $a > 0$, then $-|a| = -a < a = |a|$. If $a < 0$, then $-|a| = a < -a = |a|$.

(c) If $ab > 0$, then $|ab| = ab$. Also, a and b must have the same sign; so either both are positive, in which case $|a| = a$, $|b| = b$, and hence $|a||b| = ab$; or both are negative, in which case $|a| = -a$, $|b| = -b$, and thus $|a||b| = (-a)(-b) = ab$. If $ab < 0$, then $|ab| = -(ab)$. Also, a and b must have opposite signs. Without loss of generality, we may assume that $a > 0$ and $b < 0$, for otherwise we could interchange the roles of a and b. Then $|a| = a$, and $|b| = -b$, so $|a||b| = a(-b) = -(ab)$. Thus for every choice of a and b, $|ab| = |a||b|$.

(d) Follows immediately from (c) with $b = a$, since $|a^2| = a^2$.

(e) Follows immediately from (b) and (c).

(f) Left as an exercise.

(g) We shall use the contrapositive technique of proof here. Suppose that (g) is not true; i.e., suppose that $|a| + |b| < |a + b|$ for some choice of a and b. Then since each side of this inequality is nonnegative, we can square each side to obtain.

$$|a|^2 + 2|a||b| + |b|^2 < |a + b|^2$$

which can be simplified by (d) to

$$a^2 + 2|a||b| + b^2 < (a + b)^2$$

or

$$a^2 + 2|a||b| + b^2 < a^2 + 2ab + b^2$$

which reduces to $|a||b| < ab$. But this is the negation of (e).

(h) Write $a - b = a + (-b)$. Then replace b in (g) by $-b$ and use (a).

(i) Since $a = a - b + b$, $|a| = |a - b + b| \le |a - b| + |b|$ by (g). Hence

$$|a| - |b| \le |a - b| \tag{1}$$

Also, since $b = b - a + a$, $|b| = |b - a + a| \le |b - a| + |a|$ by (g), and so $|b| - |a| \le |b - a|$. But $|b - a| = |a - b|$ by (a). Thus we have $|b| - |a| \le |a - b|$ or, multiplying through by -1,

$$|a| - |b| \ge -|a - b| \tag{2}$$

Combining (1) and (2) yields $-|a - b| \le |a| - |b| \le |a - b|$. By Lemma 3.9, with $x = |a| - |b|$ and $c = |a - b|$, we obtain

$$||a| - |b|| \le |a - b|$$

which is the desired result.   □

With the use of absolute value, we can now establish a useful criterion for boundedness of a set.

Theorem 3.13

A set A is bounded iff there exists a nonnegative number M such that $|x| \leqslant M$ for every $x \in A$.

Proof We show first that if $|x| \leqslant M$ for every $x \in A$, then A is bounded. By Lemma 3.9, $-M \leqslant x \leqslant M$ for every $x \in A$. Hence M is an upper bound for A and $-M$ is a lower bound for A, and so A is bounded. To complete the proof, we must show that if A is bounded, there exists M such that $|x| \leqslant M$ for every $x \in A$. Since A is bounded, A must be bounded both above and below. Let M_1 be an upper bound for A and M_2 be a lower bound for A; that is, $M_2 \leqslant x \leqslant M_1$ for every $x \in A$. Let M be the larger of the two numbers $|M_1|$ and $|M_2|$ or their common value if they are equal, which we normally indicate by writing $M = \max \{|M_1|, |M_2|\}$. Then, $-M \leqslant x \leqslant M$ for every $x \in A$, which by Lemma 3.9 says that $|x| \leqslant M$ for every $x \in A$. $\square$

PROBLEMS

3.1 For a given nonempty set A, define the set B by $B = \{x | -x \in A\}$. Prove:
(a) c is a lower bound for A iff $-c$ is an upper bound for B.
(b) d is an upper bound for A iff $-d$ is a lower bound for B.
(c) Conclude that A is bounded iff B is bounded.

3.2 Show that a set cannot have two distinct upper bounds both of which belong to the set.

3.3 Show that for $c \geqslant 0$, $\{x \mid |x| > c\} = \{x \mid x > c\} \cup \{x \mid x < -c\}$.

3.4 Express each set A in terms of inequalities involving x, with no absolute-value symbols:

(a)
$$A = \{x \mid |x - 2| \leqslant 1\}$$

(b)
$$A = \{x \mid 0 < |x - 2| \leqslant 1\}$$

(c)
$$A = \{x \mid |x - 3| \geqslant 2\}$$

3.5 Complete the proofs of (d), (e), and (h) of Theorem 3.12.

3.6 Prove (f) of Theorem 3.12.

3.7 Carry out an alternate proof of (g), Theorem 3.12, in the following way. Use Theorem 3.12b, on a and b separately, add the resulting inequalities, then apply Lemma 3.9.

3.8 Show that if $|x - 2| < 1$, then $|(x - 2)(x + 5)| < 8$.

3.9 Prove that a nonempty set A in R_1 is bounded iff there exists a real number $K \geqslant 0$ such that $|x - y| \leqslant K$ for every x, y in A.

3.10 Prove that the union of two bounded sets is bounded.

3.11 Prove that for any positive integer n, the union of n bounded sets is bounded.

3.12 If a and b are real numbers with $a < b$, prove that

$$\{x \mid a < x < b\} = \left\{x \mid \left|x - \frac{a+b}{2}\right| < \frac{b-a}{2}\right\}$$

3.13 Express each set A in terms of a single inequality involving absolute-value symbols:

(a) $\qquad\qquad\qquad\qquad A = \{x \mid 2 < x < 8\}$

(b) $\qquad\qquad\qquad\qquad A = \{x \mid -2 \leqslant x \leqslant 3\}$

(c) $\qquad\qquad\qquad\qquad A = \{x \mid 0 < x < 4\}$

3.14 If a and b are real numbers with $a < b$, prove that

$$\{x \mid x < a\} \cup \{x \mid x > b\} = \left\{x \mid \left|x - \frac{a+b}{2}\right| > \frac{b-a}{2}\right\}$$

3.15 Express each set A in terms of a single inequality involving absolute value symbols:

(a) $\qquad\qquad\qquad A = \{x \mid x < -1\} \cup \{x \mid x > 3\}$

(b) $\qquad\qquad\qquad A = \{x \mid x \leqslant 5\} \cup \{x \mid x \geqslant 6\}$

4
Suprema and Infima; Finite and Infinite Sets

From our discussion of bounded sets of real numbers, it is evident that if b is an upper bound for a set A, then every real number $c > b$ is also an upper bound for A. Thus a set A which is bounded above has no largest upper bound. In this section we consider the following deeper question: Does a set A which is bounded above have a smallest upper bound? If so, we call such a number a supremum of the set or a least upper bound for the set.

Definition 4.1

Let A be a subset of R_1. A real number b is said to be a *supremum* of A (written $b = \sup A$) or a *least upper bound* for A iff both the following conditions are satisfied:

(a) b is an upper bound for A.
(b) If $c < b$, then c is not an upper bound for A.

It is sometimes convenient to use condition (b) in its contrapositive form, which we list for easy reference.

(b')   If c is an upper bound for A, then $c \geqslant b$.

By virtue of condition (b), *the empty set $\varnothing$ has no supremum in* R_1 since every real number is an upper bound for $\varnothing$ according to Corollary 3.4.

Completeness axiom (or least upper bound axiom) for R_1

The ordered field R_1 is *complete;* i.e., every nonempty subset of R_1 which is bounded above has a supremum.

We first show that a supremum of a nonempty set, when it exists, is unique.

Lemma 4.2

A nonempty set in R_1 has at most one supremum.

Proof   Let A be a nonempty set in R_1 and suppose that $x = \sup A$ and $y = \sup A$.   By Definition 4.1a, x and y are both upper bounds for A.   Then by (b'), $x = \sup A$ implies $y \geqslant x$, and $y = \sup A$ implies $x \geqslant y$.   Therefore $x = y$. $\square$

Henceforth we may refer to *the* supremum of a nonempty set which is bounded above, since its existence is guaranteed by the least upper bound axiom and its uniqueness is assured by Lemma 4.2.   We are now able to prove that among all the lower bounds for a nonempty set, there is a largest one, which we call its infimum.

Definition 4.3

Let A be a subset of R_1.   A real number b is said to be an *infimum* of A (written $b = \inf A$) or a *greatest lower bound* for A iff both the following conditions are satisfied:

(a) b is a lower bound for A.
(b)   If $c > b$, then c is not a lower bound for A.

Since every real number is a lower bound for $\varnothing$ by Corollary 3.4, it follows from condition (b) that *the empty set $\varnothing$ has no infimum in* R_1.   We leave as exercises the statement of the contrapositive form of condition (b) and the proof that a nonempty set has at most one infimum.

Theorem 4.4

Every nonempty set of real numbers which is bounded below has an infimum.

Proof Let H be a nonempty set of real numbers which is bounded below, and let b be any lower bound for H. Define the set K as follows: $K = \{p \mid -p \in H\}$. Then K is nonempty, and $-b$ is an upper bound for K by Prob. 3.1. By the least upper bound axiom, K has a supremum, which we denote by $-c$. We shall show that $c = \inf H$. That c is a lower bound for H follows from Prob. 3.1 and the fact that $-c$ is an upper bound for K. Now suppose $d > c$. Then $-d < -c$, and since $-c = \sup K$, we see that $-d$ is not an upper bound for K. Hence by Prob. 3.1, d is not a lower bound for H.   □

Combining the least upper bound axiom and Theorem 4.4 yields the following.

Theorem 4.5

Every bounded nonempty set of real numbers has both a supremum and an infimum. Furthermore, for any such set A we have $\inf A \leqslant \sup A$, with equality holding iff A is a singleton.

We emphasize that Theorem 4.5 asserts the *existence* of the numbers $\inf A$ and $\sup A$ for a bounded nonempty set A in R_1 but makes no assertion about whether or not these numbers are members of A. In the case where $\inf A \in A$ and $\sup A \in A$, we shall say that the set A *contains* both its supremum and its infimum. The reader may verify that for the set $A = \{1,2,3\}$ of Example 3.2, we have $\inf A = 1$ and $\sup A = 3$, so that the set A contains both its supremum and its infimum. However, for the set $A = \{x \mid 0 < x < 1\}$ of Example 3.5, we have $\inf A = 0$ and $\sup A = 1$, and in this case the set A contains neither its infimum nor its supremum. We shall use these notions to define finite and infinite sets.

Before doing so, we remark that there are various equivalent formulations of the terms "finite" and "infinite." Two commonly used formulations are as follows.

A set H is *infinite* iff there is a one-to-one correspondence between H and a proper subset of H. A set H is *finite* iff H is not infinite.

A nonempty (the empty set is defined to be finite) set H is finite iff there exists a positive integer n and a one-to-one correspondence between H and the set $I_n = \{1,2, \ldots ,n\}$. A set H is infinite iff H is not finite.

The usual procedure is to select one of the above as the definition of the terms "finite set" and "infinite set" and then prove the other as a theorem. However, both formulations involve the notion of a one-to-one correspondence, which is a special kind of mapping discussed in Part Two. We therefore choose yet another formulation, which depends only on concepts already discussed. Its equivalence with the above assertions is established in Part Two, where the latter are proved as theorems (see Theorems 13.16 and 13.18).

Definition 4.6

The empty set $\varnothing$ is said to be *finite*. A nonempty set H in R_1 is said to be *finite* iff H is bounded and every nonempty subset K of H contains both inf K and sup K.

A set which is not finite is called infinite and is defined formally using the contrapositive of Definition 4.6.

Definition 4.7

A set H in R_1 is said to be *infinite* iff either

(a) H is unbounded, or
(b) H is bounded and contains a nonempty subset K such that inf $K \notin K$ or sup $K \notin K$.

It should be evident that every superset of an infinite set is infinite and every subset of a finite set is finite. Examples of infinite sets are R_1 and I, as well as the set $A = \{x \mid 0 < x < 1\}$ of Example 3.5. Note that R_1 and I are both unbounded, whereas the set A is bounded but contains neither inf A nor sup A. A seemingly trivial but very useful example of a finite set is a singleton. For if $H = \{p\}$, then the only nonempty subset of H is H itself, and we obviously have inf $H =$ sup $H = p \in H$. We state this result as a theorem for easy reference.

Theorem 4.8

A singleton is a finite set.

Theorem 4.9

The union of two finite sets is a finite set.

Proof Let A and B be finite sets, and let K be any nonempty subset of

$A \cup B$. Since the set $A \cup B$ is bounded, by Prob. 3.10, our proof will be complete if we can show that inf $K \in K$ and sup $K \in K$. Let $H_1 = K \cap A$ and $H_2 = K \cap B$, noting that $K = H_1 \cup H_2$. If $H_1 = \varnothing$, then $K = H_2$ is a nonempty subset of B and hence is finite, and so inf $K \in K$ and sup $K \in K$, with a similar result holding if $H_2 = \varnothing$. We may thus suppose that H_1 and H_2 are nonempty. Since H_1 is a nonempty subset of the finite set A, we know that inf $H_1 \in H_1$ and sup $H_1 \in H_1$. Similarly, H_2 is a nonempty subset of B, so that inf $H_2 \in H_2$ and sup $H_2 \in H_2$. We leave as a simple exercise the proof that inf K is the minimum of the two numbers inf H_1 and inf H_2, and hence that inf $K \in K$; also that sup K is the maximum of the two numbers sup H_1 and sup H_2, and hence that sup $K \in K$. $\square$

Using Theorem 4.9 and mathematical induction, the following corollary is immediate, and its proof is left as an exercise.

Corollary 4.10

For any positive integer n, let $A_1, A_2, \ldots, A_n$ be finite sets. Then $A_1 \cup A_2 \cup \cdots \cup A_n$ is a finite set.

Theorem 4.11

If H is a nonempty finite set of real numbers, then there exists a positive integer n such that $H = \{a_1, a_2, \ldots, a_n\}$, where $a_1 < a_2 < \cdots < a_n$.

Proof Let H be a nonempty finite set, let $a_1 = $ inf H, and define $H_1 = H - \{a_1\}$. If $H_1 = \varnothing$, then $H = \{a_1\}$ and we are finished. If $H_1 \neq \varnothing$, let $a_2 = $ inf H_1 and define $H_2 = H_1 - \{a_2\}$. If $H_2 = \varnothing$, then $H = \{a_1, a_2\}$ and we are finished. If $H_2 \neq \varnothing$, let $a_3 = $ inf H_2 and define $H_3 = H_2 - \{a_3\}$. This process can clearly be continued by mathematical induction, since if $H_k \neq \varnothing$ for $k \in I$, we let $a_{k+1} = $ inf H_k and define $H_{k+1} = H_k - \{a_{k+1}\}$. We shall now show that for some $n \in I$ we must have $H_n = \varnothing$, and hence $H = \{a_1, a_2, \ldots, a_n\}$. Let us assume by way of contradiction that $H_n \neq \varnothing$ for every $n \in I$. Then the set $K = \{a_n | n \in I\}$, where $a_1 < a_2 < a_3 < \cdots$, is a subset of H and hence sup $K \in K$. Thus we must have sup $K = a_k$ for some $k \in I$. But this is impossible, since $a_k < a_{k+1}$ and $a_{k+1} \in K$. $\square$

Theorem 4.12

Let H be a bounded nonempty set of real numbers, U the set of all upper bounds for H, and L the set of all lower bounds for H. Then:

(a) U has an infimum, and inf $U \in U$.

(b) L has a supremum, and sup $L \in L$.

Proof It follows from Theorem 4.5 that H has both an infimum a and a supremum b. We shall prove (a) and leave (b) as an exercise. Clearly, b is an upper bound for H, and hence $b \in U$. Our proof will be completed by showing that $b = \inf U$, so let x be any member of U. Then x is an upper bound for H, and since $b = \sup H$, we must have $x \geqslant b$. Hence b is a lower bound for U. Since $b \in U$, it is evident that no real number $c > b$ can be a lower bound for U. Therefore $b = \inf U$. ☐

Another important application of the least upper bound axiom is in the proof of the following theorem.

Theorem 4.13   Archimedean property of R_1

Given any $\epsilon > 0$, there exists a positive integer N such that $N\epsilon > 1$.

Proof Suppose the theorem is not true. Then for some $\epsilon > 0$ we must have $n\epsilon \leqslant 1$ for every positive integer n; that is, the set $H = \{\epsilon, 2\epsilon, 3\epsilon, \ldots\}$ is bounded above. By the least upper bound axiom, H must have a supremum, which we denote by c, and hence $n\epsilon \leqslant c$ for every $n \in I$. In particular, for each $n \in I$, we must have $(n+1)\epsilon \leqslant c$, from which it follows that $n\epsilon \leqslant c - \epsilon$. But this last inequality asserts that $c - \epsilon$ is an upper bound for H, which is impossible since $c - \epsilon < c$ and $c = \sup H$. ☐

Corollary 4.14

For any real number x, there exists a positive integer N such that $N > x$.

Proof If $x \leqslant 0$, we may choose N as any positive integer, say $N = 1$. If $x > 0$, then $1/x > 0$ and, by Theorem 4.13, there exists a positive integer N such that $N(1/x) > 1$, or $N > x$. ☐

Corollary 4.15

If a and b are positive numbers, there exists a positive integer N such that $Na > b$.

Proof Let $x = b/a$. By Corollary 4.14, there exists a positive integer N such that $N > x$. Thus, $N > b/a$, and since a is positive, $Na > b$. ☐

Note carefully that if in Corollary 4.15 we choose $a = \epsilon$ and $b = 1$, we then obtain Theorem 4.13 as a special case of this corollary. Thus Corollary 4.15 implies Theorem 4.13. It follows that the three properties stated in the theorem and its corollaries are all equivalent expressions for the Archimedean property of R_1.

PROBLEMS

4.1 State the contrapositive of condition (b) in Definition 4.3, and prove that a nonempty set has at most one infimum.

4.2 Use Definition 4.6 to show that the set $A = \{1,2,3\}$ is finite. Give an alternate proof that A is finite using Theorem 4.8 and Corollary 4.10.

4.3 If we define the set I_n for each $n \in I$ by $I_n = \{1,2,3, \ldots ,n\}$, prove that for any positive integer n, the set I_n is finite and the set $I - I_n$ is infinite.

4.4 Complete the proof of Theorem 4.9.

4.5 Prove Corollary 4.10.

4.6 Prove (b) of Theorem 4.12.

4.7 Under what condition could we have sup L = inf U in Theorem 4.12?

4.8 If $A = \{x|0 < x < 1\}$, prove that inf $A = 0$ and sup $A = 1$.

4.9 Given the set

$$H = \left\{ \frac{2}{3}, \frac{4}{5}, \frac{6}{7}, \frac{8}{9}, \; \cdots \; , \frac{2n}{2n+1}, \; \cdots \right\}$$

(a) Show that H is bounded.
(b) Find inf H and sup H.
(c) Determine a member of H which is greater than sup $H - \frac{1}{100}$.
(d) Prove that H is infinite.

4.10 Let A be a nonempty set, and define B by $B = \{x|-x \in A\}$.
(a) Prove that $c = \inf A$ iff $-c = \sup B$.
(b) Prove that $d = \sup A$ iff $-d = \inf B$.
(c) Prove that A is finite iff B is finite.

4.11 If A is a nonempty subset of a bounded set B of real numbers, prove that

$$\inf B \leqslant \inf A \leqslant \sup A \leqslant \sup B.$$

4.12 Give a direct proof that Corollary 4.15 implies Corollary 4.14.

4.13 Prove that if $b \in H$ and if b is an upper bound for H, then $b = \sup H$.

4.14 Prove that if $b \in H$ and if b is a lower bound for H, then $b = \inf H$.

4.15 Let A and B be sets in R_1 with the property that sup $A <$ inf B. Prove that

$$A \cap B = \varnothing.$$

4.16 Prove that for any two real numbers a and b,

$$\sup \{a,b\} = \frac{a + b + |a - b|}{2}$$

and

$$\inf \{a,b\} = \frac{a + b - |a - b|}{2}$$

4.17 If a, b, c, d are real numbers, show that

$$\sup \{a + c, \, b + d\} \leq \sup \{a,b\} + \sup \{c,d\}$$

4.18 If H is a bounded nonempty set of real numbers, show that the nonnegative real number

$$d(H) = \sup \{|x - y| \mid x \in H, \, y \in H\}$$

exists. (*Hint:* See Prob. 3.9.) The nonnegative real number $d(H)$ is called the *diameter* of the bounded set H.

4.19 (*a*) Let H be a bounded set of real numbers containing at least two points. Prove that $b = \sup H$ iff given any $\epsilon > 0$ the following two conditions are satisfied:

(i) $x < b + \epsilon$ for every $x \in H$

(ii) $x > b - \epsilon$ for at least one $x \in H$

(*b*) With H as in part (*a*), state and prove a similar characterization for $a = \inf H$.

5
Connected Sets in R_1; Intervals and Rays; Separated Sets

Among all the subsets of R_1, there are certain types of sets which play a vital role in analysis. We begin our study of such sets in this chapter with the introduction of the class of connected sets. Bear in mind that our universe is the set R_1 of real numbers, or equivalently, the set of points on the real line, and that we shall use the terms "real number" and "point" interchangeably.

Definition 5.1

A point p is said to be *between* a and b iff either $a < p < b$ or $b < p < a$. A subset H of R_1 is *connected* iff given any two points a, b of H, every point between a and b is a point of H.

Since the negation of Definition 5.1 will be a useful tool throughout this chapter, we list it as a theorem.

Theorem 5.2

A subset H of R_1 is not connected iff there exist two points a, b of H and a point p between a and b such that $p \notin H$.

By virtue of this theorem, we see that in order for a set not to be connected, the set must contain at least two points. Thus the following result is immediate.

Corollary 5.3

The empty set $\varnothing$ is connected, and every singleton set is connected.

Theorem 5.4

The intersection of two connected sets in R_1 is connected.

Proof By virtue of Corollary 5.3, it suffices to consider connected sets A and B for which $A \cap B$ contains at least two points. The result is then immediate from Definition 5.1 and the relations $A \cap B \subset A$ and $A \cap B \subset B$.   □

Lemma 5.5

If A is a nonempty connected subset of R_1 and $p \notin A$, then exactly one of the following two statements is true:

(a) $p < x$ for every $x \in A$

(b) $p > x$ for every $x \in A$

Proof Let $a \in A$. Clearly, $a \neq p$, and so we must have $a < p$ or $a > p$ but not both. The result then follows from either Definition 5.1 or Theorem 5.2.   □

Theorem 5.6

If the intersection of two connected sets is nonempty, their union is connected.

Proof Let A and B be connected sets with $A \cap B \neq \varnothing$, and assume that $A \cup B$ is not connected. By Theorem 5.2, there exist points $a, b \in A \cup B$ and a point p between a and b such that $p \notin A \cup B$. We may choose our notation so that $a \in A$ and $a < p < b$. Then clearly, we must have $b \in B$. By Lemma 5.5, it follows that $x < p$ for every $x \in A$ and $x > p$ for every $x \in B$, and hence $A \cap B = \varnothing$, a contradiction.   □

Table 5.1 lists in convenient catalog form the subsets (including appropriate terminology and notation) of R_1 which are connected.

It is an easy exercise to verify that all sets in the catalog are connected, and we now show that every connected subset of R_1 is in the catalog.

Table 5.1   **Catalog of connected sets in R_1**

1. The empty set $\varnothing$
2. Singletons $\{p\}$
3. Open intervals $(a,b) = \{x \mid a < x < b\}$
4. Closed intervals $[a,b] = \{x \mid a \leqslant x \leqslant b\}$ $\quad$ where $a,b \in R_1$
5. Half-open (or half-closed) intervals $\quad$ with $a < b$
   $(a,b] = \{x \mid a < x \leqslant b\}$ $\quad$ and $\quad$ $[a,b) = \{x \mid a \leqslant x < b\}$
6. Open rays $\{x \mid a < x\}$ and $\{x \mid x < a\}$ $\quad$ where $a \in R_1$
7. Closed rays $\{x \mid a \leqslant x\}$ and $\{x \mid x \leqslant a\}$
8. The set R_1

Theorem 5.7

A subset H of R_1 is connected iff H is in the catalog.

Proof In view of Corollary 5.3, it suffices to show that if H is a connected subset of R_1 which contains at least two points, then H is in the catalog. Since such a set H is either bounded or unbounded, we consider separate cases.

Case 1 H is bounded. Let $a = \inf H$ and $b = \sup H$, noting that $a < b$. We shall show that H is one of the intervals of type 3, 4, or 5 in Table 5.1. For suppose the contrary. Then there exists a point p between a and b such that $p \notin H$. Then by Lemma 5.5, either $p \quad x$ for every $x \in H$, or $p > x$ for every $x \in H$. But the first inequality contradicts our choice of a and the second contradicts our choice of b.

Case 2 H is bounded above but not bounded below. Let $a = \sup H$. We leave as an exercise the proof that H is the open ray $\{x \mid x < a\}$ or the closed ray $\{x \mid x \leqslant a\}$.

Case 3 H is bounded below but not bounded above. Let $a = \inf H$. Then H is the open ray $\{x \mid x > a\}$ or the closed ray $\{x \mid x \geqslant a\}$.

Case 4 H is neither bounded above nor bounded below. Then $H = R_1$. $\quad\square$

For any of the intervals 3 to 5 in Table 5.1, a and b are called the *end points* of the interval, and the real number $b - a > 0$ is called the *length* of the interval. Occasionally, we regard a singleton $\{p\}$ as a closed interval with end points $a = b = p$. Such a closed interval has length zero and is said to be *degenerate*.

An open interval is clearly an infinite set, since it contains neither its supremum nor its infimum.

Corollary 5.8

Every connected subset of R_1 containing more than one point is an infinite set.

Proof Every such set is a superset of an open interval.   □

Suppose now that H is a set which is not connected. We know there exist two points $a,b \in H$ and a point p between a and b such that $p \notin H$. We define

$$A_p = \{x \mid x \in H \text{ and } x < p\} \qquad B_p = \{x \mid x \in H \text{ and } x > p\}$$

noting that A_p and B_p are disjoint nonempty sets and that

$$H = A_p \cup B_p$$

Let q be any point of A_p. Then the open interval $(q - 1, p)$ contains no point of B_p. For this reason, we say that q is not a cluster point of B_p in accordance with the following definitions.

Definition 5.9

If (a,b) is an open interval in R_1 and $p \in (a,b)$, then the set $(a,b) - \{p\}$ is called a *deleted open interval about p*. For convenience of notation, we shall denote such a deleted open interval as (a,p,b) with the understanding that

$$(a,p,b) = (a,b) - \{p\} = (a,p) \cup (p,b)$$

Definition 5.10

Let A be a subset of R_1. Then a point $p \in R_1$ is called a *cluster point* of A iff every deleted open interval about p contains at least one point of A.

Returning to our discussion of the sets A_p and B_p, we see that no point of A_p is a cluster point of B_p, and the reader may verify that no point of B_p is a cluster point of A_p. We say that A_p and B_p are separated sets.

Definition 5.11

Two sets A and B are said to be *separated* iff they are disjoint and nonempty and neither contains a cluster point of the other.

Note that although the sets $A = (0,1)$ and $B = [1,2)$ are nonempty and disjoint, they are not separated, since 1 is a point of B which is a cluster point of A. However, the sets A and $C = (1,2)$ are separated.

It is noteworthy that $A \cup B$ is the open interval $(0,2)$, which is a connected set, whereas $A \cup C$ is the deleted open interval $(0,1,2)$, which is not connected. This observation leads to one of the principal characterizations of connected sets.

Theorem 5.12

A set H is connected iff it is not the union of two separated sets.

Proof We have shown that if H is not connected, then H is the union of two separated sets. For the converse, suppose that H is the union of two separated sets A and B. Let us choose our notation so that $a \in A$, $b \in B$, and $a < b$. Define the set $K = \{x \in A \mid x < b\}$. Clearly K is nonempty since $a \in K$, and b is an upper bound for K. Defining $p = \sup K$, we see at once that $a \leq p \leq b$. Now let us assume by way of contradiction that H is connected. Then $[a,b] \subset H$, and hence $p \in H$, so that p is in exactly one of the sets A, B. We suppose $p \in A$, leaving the similar case $p \in B$ as an exercise. Then $p \notin B$, so that $p < b$, and hence $(p,b] \subset B$. It is easy to see that p is a cluster point of the half-open interval $(p,b]$ and hence of its superset B, which contradicts the hypothesis that A and B are separated. $\square$

Corollary 5.13

Let H be a connected set which is the union of two disjoint nonempty sets. Then at least one of these sets must contain a cluster point of the other.

Corollary 5.14

The set R_1 is not the union of two separated sets. That is, if A and B are disjoint nonempty sets whose union is R_1, then at least one of the sets A, B must contain a cluster point of the other.

Although the proof of Corollary 5.14 depends heavily upon the least upper bound (or completeness) axiom, it is also a fact that this corollary implies the completeness axiom, so that these two statements are equivalent. In other words, by considerably rearranging our work, we could have chosen the statement of Corollary 5.14 as an axiom and then proved the completeness property of R_1 as a theorem.

Theorem 5.15

If we accept as true the statement of Corollary 5.14, then every nonempty set H which is bounded above has a supremum.

Proof Let us assume by way of contradiction that the set H has no
supremum. Let a be any point of H and b any upper bound of H.
Then R_1 is clearly the union of the disjoint sets G_1 and G_2 defined
by

$$G_1 = \{x \in R_1 \mid x \text{ is not an upper bound of } H\}$$
$$G_2 = \{x \in R_1 \mid x \text{ is an upper bound of } H\}$$

Each set is nonempty since $a - 1 \in G_1$ and $b \in G_2$. By Corollary 5.14, there is a point p in one of these sets which is a cluster
point of the other set.

Suppose $p \in G_1$. Then there is a point $s \in H$ such that $p < s$,
and the deleted open interval $(p - 1, p, s)$ contains no point of
G_2, contradicting the fact that p is a cluster point of G_2.

Suppose $p \in G_2$. Then p is an upper bound of H, but p is not
the least upper bound for H. Thus there is a point $r < p$ such
that r is an upper bound of H, and $(r, p, p + 1)$ contains no point
of G_1, contradicting the fact that p is a cluster point of G_1. $\square$

PROBLEMS

5.1 Complete the proof of Theorem 5.7.

5.2 We may characterize an open interval as a nonempty bounded connected subset of
R_1 which contains neither its infimum nor its supremum. Determine similar characterizations for the other types of intervals and rays.

5.3 Let (a,b) and (c,d) be open intervals such that $(a,b) \cap (c,d) \neq \emptyset$. Prove that the
points of the set $\{a, b, c, d\}$ can be renamed a_1, a_2, a_3, a_4, in such a way that

$$a_1 \leq a_2 \leq a_3 \leq a_4, \text{ where } a_1 \neq a_3 \text{ and } a_2 \neq a_4.$$

(*Hint:* Define $a_1 = \inf \{a,c\}$, etc.)   Conclude that the following assertions are true:
  (a) $(a,b) \cup (c,d) = (a_1, a_4)$.
  (b) $(a,b) \cap (c,d) = (a_2, a_3)$.
  (c) If $[r,s]$ is any closed interval such that $[r,s] \subset (a,b) \cup (c,d)$, then

$$a_4 - a_1 > s - r.$$

  (d) $(b - a) + (d - c) > s - r$.

5.4 Use Prob. 5.3 to determine whether the following statements are theorems:
  (a) The intersection of two open intervals is either the empty set or an open interval.
  (b) If the intersection of two open intervals is nonempty, then their union is an
open interval; furthermore, the length of their union is less than the sum of the lengths
of the given intervals.

5.5 Let n be a positive integer, and for each $i = 1, 2, \ldots, n$, let A_i be a subset of R_1.
Define $A = A_1 \cap A_2 \cap \cdots \cap A_n$ and $H = A_1 \cup A_2 \cup \cdots \cup A_n$. Prove that:
  (a) If each A_i is connected, $i = 1, \ldots, n$, then A is connected. (This generalizes
Theorem 5.4 to an arbitrary finite number of sets.)
  (b) If each A_i is connected and $A \neq \emptyset$, then H is connected. (This generalizes
Theorem 5.6 to an arbitrary finite number of sets.)

(c) If each A_i is an open interval, then A is either the empty set or an open interval.

(d) If each A_i is an open interval and $A \neq \varnothing$, then H is an open interval.

5.6 Is the intersection of two rays necessarily a ray? Explain.

5.7 If H is any interval (open, closed, or half-open) with end points $a < b$, show that $b - a = d(H)$, the diameter of H (see Prob. 4.18).

5.8 Complete the proof of Theorem 5.12 by considering the case $p \in B$.

5.9 Show that R_1 has the *Hausdorff property:* For any points $p, q \in R_1$ with $p \neq q$, there exist disjoint open intervals (a,b) and (c,d) such that $p \in (a,b)$ and $q \in (c,d)$.

5.10 Prove that if H is a nonempty connected set and $H \subset A \cup B$, where A and B are separated, then H is contained in exactly one of the sets A, B.

6

The Heine-Borel and Bolzano-Weierstrass Theorems

We start this chapter with some further results on cluster points.

Definition 6.1

Let A be a nonempty subset of R_1. Then:

(a) We denote by A' the set of all cluster points of A and call A' the *derived set* of A.

(b) A point $p \in A$ is called an *isolated point* of A iff there exists a deleted open interval (a,p,b) such that $(a,p,b) \cap A = \varnothing$.

(c) We call A an *isolated set* iff every point of A is isolated.

Clearly, if A is any nonempty subset of R_1, then each point of A is either an isolated point or a cluster point of A. Note that every isolated point of A must be a member of A, whereas a cluster point of A may (but need not) be a member of A. For example, if A is the open interval (a,b), then every point of A is a cluster point of A, and both end points a and b are cluster points of A which are not members of A.

Thus if $A = (a,b)$, then $A' = [a,b]$. Our next two theorems generalize this example in two different directions. The first is an immediate consequence of our definitions of suprema and infima, and the second follows from our work with connected sets.

Theorem 6.2

Let A be a nonempty subset of R_1. Then:

(a) If A is bounded above and sup $A \notin A$, then sup $A \in A'$.
(b) If A is bounded below and inf $A \notin A$, then inf $A \in A'$.

Theorem 6.3

If A is a connected subset of R_1 containing more than one point, then $A \subset A'$.

Theorem 6.4

Let A and B be sets in R_1. Then

(a)   If $A \subset B$, then $A' \subset B'$
(b) $(A \cup B)' = A' \cup B'$
(c) $(A \cap B)' \subset A' \cap B'$
(d) $(A')' \subset A'$
(e) $\varnothing' = \varnothing$
(f) $\{p\}' = \varnothing$
(g) $R_1' = R_1$

Proof It is clear from Definition 5.10 that if p is a cluster point of A, then p is a cluster point of every superset of A, which proves (a). For (b), let $H = A \cup B$. Since $A \subset H$ and $B \subset H$, it follows from (a) that $A' \subset H'$ and $B' \subset H'$, and hence $A' \cup B' \subset H'$. For the reverse inclusion we prove the contrapositive; let us suppose that $p \notin A' \cup B'$. Since $p \notin A'$, there exists a deleted open interval (a,p,b) which contains no point of A, and since $p \notin B'$, there exists a deleted open interval (c,p,d) which contains no point of B. Letting $r = \sup \{a,c\}$ and $s = \inf \{b,d\}$, we see that the deleted open interval (r,p,s) contains no point of H, and so $p \notin H'$. This proves (b). The proof of (c) follows from (a) since $A \cap B \subset A$ and $A \cap B \subset B$. The proofs of (d) to (g) are left as exercises.   □

Since an isolated set consists entirely of isolated points, it may seem surprising that an isolated set may have a cluster point, as shown in our next example. Of course, in this case the cluster point cannot be a member of the set.

Example 6.5

Let $H = \{1/n \mid n \in I\}$. It is easy to show that H is an isolated set, for if $p \in H$, then $p = 1/n$ for some $n \in I$. The reader may verify that the deleted open interval $(p - 1/(n + 1),\ p,\ p + 1/(n + 1))$ contains no point of H, so that p is an isolated point of H. On the other hand, the reader may also verify that $0 = \inf H$, and since $0 \notin H$, it follows from Theorem 6.2b that $0 \in H'$.

The situation illustrated in Example 6.5 cannot happen when the points of an isolated set are "sufficiently" isolated, in the sense of the following definition.

Definition 6.6

A set H in R_1 is said to be *uniformly isolated* iff there exists a real number $r > 0$ such that $|c - d| \geq r$ for any two distinct points c and d of H.

Theorem 6.7

A uniformly isolated set of real numbers has no cluster point; that is, if H is uniformly isolated, then $H' = \varnothing$.

Proof We shall show that no real number p can be a cluster point of H. Thus let p be any real number. Since H is uniformly isolated, there exists a real number $r > 0$ such that $|c - d| \geq r$ for any two distinct points c and d of H. Then in particular, the open interval $A = (p - r/2,\ p + r/2)$ can contain at most one point of H. If A contains no point of H, or if $p \in H$, then the deleted open interval $(p - r/2,\ p,\ p + r/2)$ contains no point of H. If A contains a point $h \in H$ where $h \neq p$, then h is in exactly one of the open intervals $(p - r/2,\ p)$ and $(p,\ p + r/2)$. Since both cases are similar, we suppose the former and define $s = \inf\{h - p + r/2,\ p - h\}$, from which it follows that the deleted open interval $(p - s,\ p,\ p + s)$ contains no point of H. Therefore $p \notin H'$. $\square$

The converse of Theorem 6.7 is not true. For an example of a set H which is not uniformly isolated but for which $H' = \varnothing$ see Prob. 6.9.

Theorem 6.8

If $H' \neq \varnothing$, then H is an infinite set.

Proof Since every unbounded set is infinite, we may suppose that H is bounded and let p be a cluster point of H. Defining H_1 and H_2 as in Prob. 610, we see that $p \in H_1' \cup H_2'$. If $p \in H_1'$, it is easy to verify that $p = \sup H_1$, and since $p \notin H_1$ and $H_1 \subset H$, it follows from Definition 4.7b that H is infinite. If $p \in H_2'$, then $p = \inf H_2$, and a similar argument shows that H is infinite. $\square$

Corollary 6.9

A finite set has no cluster point.

Note that Corollary 6.9 is the contrapositive of Theorem 6.8. Our next theorem characterizes bounded infinite sets by means of cluster points. This result is one of many equivalent forms of the Bolzano-Weierstrass Theorem, one of the most important theorems of analysis.

Theorem 6.10   The Bolzano-Weierstrass theorem

A bounded set of real numbers is infinite iff it has a cluster point.

Proof If H has a cluster point, then H is infinite, by Theorem 6.8. For the converse, if H is a bounded infinite set, then H contains a nonempty subset K such that $\inf K \notin K$ or $\sup K \notin K$, by Definition 4.7b. The result is then immediate from Theorem 6.2. $\square$

Theorem 6.11   The Heine-Borel theorem

Let $[a,b]$ be a closed interval and Δ a family of open intervals such that each point of $[a,b]$ is in at least one set of the family Δ. Then there is a finite collection $G_1, G_2, \ldots, G_n$ of open intervals satisfying the following conditions:

(a) G_i is a member of Δ for $i = 1, 2, \ldots, n$.
(b) Each point of $[a,b]$ is contained in at least one of the sets G_i.

Proof We define a subset L of $[a,b]$ as follows: A point p of $[a,b]$ is in L iff there is a finite collection $G_1, G_2, \ldots, G_k$ of open intervals satisfying the following conditions:

(i) G_i is a member of Δ for each $i = 1, 2, \ldots, k$.

(ii)   Each point of $[a,p]$ is contained in at least one of the sets G_i, $1 \leqslant i \leqslant k$.

Note that L is nonempty, since $a \in L$. Since our proof will be complete if we show that $b \in L$, let us suppose by way of contradiction that $b \notin L$. By hypothesis, there is an open interval (c,d) in the family Δ such that $b \in (c,d)$. If (c,d) contains any point of L, we have an immediate contradiction to our assumption that $b \notin L$. Thus we may assume that (c,d) contains no point of L, and hence c is an upper bound for the nonempty set L.

Define $z = \sup L$, and note that $z \leqslant c < b$. Clearly, there must exist an open interval (r,s) in the family Δ such that $s < b$ and $z \in (r,s)$. Hence r is not an upper bound for L, and so (r,s) must contain at least one point of L. It then follows from the definition of the set L that $(r,s) \subset L$. But this contradicts the fact that $z = \sup L$. $\square$

PROBLEMS

6.1   Prove Theorem 6.2.

6.2   Prove Theorem 6.3.

6.3   Complete the proof of Theorem 6.4.

6.4   Show that equality does not hold in part (c) of Theorem 6.4 by giving an example where $(A \cap B)' \neq A' \cap B'$.

6.5   Verify the assertions left to the reader in Example 6.5.

6.6   State carefully what is meant by the assertion that a set H in R_1 is not uniformly isolated.

6.7   Show that the set I of positive integers is uniformly isolated.

6.8   Show that the set A defined by $A = \{n + 1/n \mid n \in I\}$ is uniformly isolated.

6.9   Let $H = A \cup I$, where A is the set defined in Prob. 6.8. Show that H is not uniformly isolated but that $H' = \varnothing$. (This example shows that the converse of Theorem 6.7 is not true.)

6.10   Let H be a subset of R_1 and $p \in R_1$. Define the sets

$$H_1 = \{x \in H \mid x < p\} \qquad H_2 = \{x \in H \mid x > p\}$$

Prove that p is a cluster point of H iff p is a cluster point of at least one of the sets H_1, H_2.

6.11   Let H be the closed interval $[0,5]$, and let Δ be the family of open intervals

$$\{(x - \tfrac{1}{2}, x + \tfrac{1}{2}) \mid x \in H\}$$

Determine a finite subfamily of Δ which satisfies the conclusions of the Heine-Borel theorem.

part two

Mappings

7
Cartesian Products; Definition and Examples of Mappings

In this chapter, we shall be concerned with particular sets whose elements are *ordered pairs* of elements from a given universe. Though we do not formally define the notion of ordered pair, the terminology suggests that it is a pair of elements from the universe and that the order in which they are chosen is important. Thus, if x and y are points in the universe, we may pair them in either of two ways, to get the ordered pair $\langle x,y \rangle$ or the ordered pair $\langle y,x \rangle$. We first define the notion of equality for ordered pairs.

Definition 7.1

If $\langle a,b \rangle$ is an ordered pair of elements of U, then a is called the *first component* of $\langle a,b \rangle$ and b is called the *second component* of $\langle a,b \rangle$. Two ordered pairs are said to be *equal* iff their first components are equal, i.e., the same element of U, and their second components are equal. Thus, $\langle a,b \rangle = \langle c,d \rangle$ iff $a = c$ and $b = d$. Note that the ordered pairs $\langle x,y \rangle$ and $\langle y,x \rangle$ are equal iff $x = y$.

Suppose now that A and B are sets in a given universe. We have previously considered several methods, e.g., unions, intersections, and complements, of constructing new sets from A and B. Utilizing the notion of ordered pair, we can construct from the sets A and B still another very important set called their Cartesian product.

Definition 7.2

If A and B are sets, then the *Cartesian product* of A and B (written $A \times B$ and read "A cross B") is the set of all ordered pairs $\langle x,y \rangle$ such that $x \in A$, $y \in B$. Symbolically,

$$A \times B = \{ \langle x,y \rangle \mid x \in A, y \in B \}$$

We now give some examples of Cartesian products.

Example 7.3

Let $U = \{1,2,3,4,5,6\}$, and let A and B be the subsets of U defined by $A = \{1,2,3,5\}$, $B = \{4,5,6\}$. The set $A \times B$ may be written as

$A \times B = \{ \langle 1,4 \rangle, \langle 1,5 \rangle, \langle 1,6 \rangle, \langle 2,4 \rangle, \langle 2,5 \rangle, \langle 2,6 \rangle,$
$\qquad \langle 3,4 \rangle, \langle 3,5 \rangle, \langle 3,6 \rangle, \langle 5,4 \rangle, \langle 5,5 \rangle, \langle 5,6 \rangle \}$

while the set $B \times A$ may be written in a similar fashion as

$B \times A = \{ \langle 4,1 \rangle, \langle 4,2 \rangle, \langle 4,3 \rangle, \langle 4,5 \rangle, \langle 5,1 \rangle, \langle 5,2 \rangle,$
$\qquad \langle 5,3 \rangle, \langle 5,5 \rangle, \langle 6,1 \rangle, \langle 6,2 \rangle, \langle 6,3 \rangle, \langle 6,5 \rangle \}$

Note that the sets $A \times B$ and $B \times A$ are not equal.

Example 7.4

Define U as in Example 7.3, and note that we have

$U \times U = \{ \langle 1,1 \rangle, \langle 1,2 \rangle, \langle 1,3 \rangle, \langle 1,4 \rangle, \langle 1,5 \rangle, \langle 1,6 \rangle,$
$\qquad \langle 2,1 \rangle, \langle 2,2 \rangle, \langle 2,3 \rangle, \langle 2,4 \rangle, \langle 2,5 \rangle, \langle 2,6 \rangle,$
$\qquad \langle 3,1 \rangle, \langle 3,2 \rangle, \langle 3,3 \rangle, \langle 3,4 \rangle, \langle 3,5 \rangle, \langle 3,6 \rangle,$
$\qquad \langle 4,1 \rangle, \langle 4,2 \rangle, \langle 4,3 \rangle, \langle 4,4 \rangle, \langle 4,5 \rangle, \langle 4,6 \rangle,$
$\qquad \langle 5,1 \rangle, \langle 5,2 \rangle, \langle 5,3 \rangle, \langle 5,4 \rangle, \langle 5,5 \rangle, \langle 5,6 \rangle,$
$\qquad \langle 6,1 \rangle, \langle 6,2 \rangle, \langle 6,3 \rangle, \langle 6,4 \rangle, \langle 6,5 \rangle, \langle 6,6 \rangle \}$

Note in particular that each of the sets $A \times B$ and $B \times A$ of Example 7.3 is a subset of $U \times U$.

Example 7.5

Let the universe be R_1, and consider the Cartesian product of R_1 with itself. We see that $R_1 \times R_1$ is the set of all ordered pairs of real numbers. Geometrically, if we "cross" the real line with itself, we get the Euclidean plane. Taking the copies of R_1 as coordinate axes and their point of intersection as the origin, we see that every point in the plane corresponds to an ordered pair of real numbers, and vice versa. This is precisely the method we use in coordinatizing the plane in analytic geometry. In fact, we generally refer to the numbers x and y as the Cartesian coordinates of the point $\langle x,y \rangle$ in $R_1 \times R_1$. The word "Cartesian" is derived from the name of the great French mathematician René Descartes (1596–1650), called the father of analytic geometry.

One of the most useful applications of Cartesian products is in the definition of a function, or mapping. These two words have precisely the same meaning and will be used interchangeably throughout this text.

Definition 7.6

If A and B are sets, and if f is a nonempty subset of $A \times B$, then f is said to be a *mapping* or *function* from A into B iff the following condition holds:

(a) Given any point x of A, there is exactly one point y of B such that $\langle x,y \rangle \in f$.

We symbolize the statement "f is a mapping from A into B" by $f: A \rightarrow B$ and call the set A the *domain* of the mapping f. We abbreviate the statement "$\langle x,y \rangle \in f$" by $f(x) = y$, where $f(x)$, read "f at x," represents the *value* of the function f at the point x. We also refer to $f(x)$ as the *image* of x under the mapping f. Note carefully that for each point x in the domain A of a mapping f, the value (or image) $f(x)$ is uniquely determined; for, according to condition (a), if $\langle x,y \rangle \in f$ and $\langle x,z \rangle \in f$, then $y = z = f(x)$. If $f: A \rightarrow B$ is a mapping and H is a nonempty subset of A, we write $f(H)$ to denote the *image of the set H under the mapping f,* where

$$f(H) = \{f(x) \mid x \in H\} = \{y \in B \mid \langle x,y \rangle \in f \text{ for some } x \in H\}$$

The set $f(A)$, called the *range* of the mapping f, will also be denoted by $\{f\}$. It should be evident that the range $\{f\}$ of a mapping $f: A \rightarrow B$ is a subset of B which may be expressed in the following equivalent forms:

(a) $\{f\} = f(A)$, the image of the domain A under the mapping f
(b) $\{f\} = \{f(x) \mid x \in A\}$, the set of all values of the mapping f
(c) $\{f\} = \{y \in B \mid \langle x,y \rangle \in f \text{ for some } x \in A\}$, the set of all second components of elements of f

Example 7.7

Let $H = \{1,2,3,4,5,6,7\}$ and $K = \{1,3,5,7,9\}$. Consider the mapping $f: H \rightarrow K$ defined as follows:

$$f = \{\langle 1,3\rangle, \langle 2,5\rangle, \langle 3,7\rangle, \langle 4,3\rangle, \langle 5,3\rangle, \langle 6,9\rangle, \langle 7,5\rangle\}$$

This mapping f can also be defined by specifying the image under f of each element of H as follows:

$$f(1) = 3 \qquad f(2) = 5 \qquad f(3) = 7$$
$$f(4) = 3 \qquad f(5) = 3 \qquad f(6) = 9 \qquad f(7) = 5$$

In this example, the domain of f is H, and the range of f is the set $\{3,5,7,9\}$, which is a proper subset of K. We note that the set $A = \{1,2\}$ is a subset of H and that $f(A) = \{3,5\}$.

Example 7.8

Let H and K be defined as in Example 7.7, and define f as follows:

$$f = \{\langle 1,3\rangle, \langle 2,3\rangle, \langle 3,3\rangle, \langle 4,3\rangle, \langle 5,3\rangle, \langle 6,3\rangle, \langle 7,3\rangle\}$$

Here the domain of f is H, and the range of f is the singleton $\{3\}$. In this case, we call f a constant mapping.

Definition 7.9

If $f: H \rightarrow K$, and if for each $x \in H$, $f(x) = k$, where k is a given element of K, then f is called a *constant mapping*.

Example 7.10

Let H and K be defined as in Example 7.7, and define f as follows:

$$f(1) = 3 \qquad f(2) = 5 \qquad f(3) = 7 \qquad f(4) = 3$$
$$f(5) = 3 \qquad f(6) = 9 \qquad f(7) = 1$$

Again, the domain of f is the set H, but the range of f is the entire set K. This illustrates the concept of an "onto" mapping.

Definition 7.11

A mapping $f: H \rightarrow K$ is said to be a mapping of H *onto* K, and we write $f: H \xrightarrow{\text{onto}} K$ iff $f(H) = K$.

We point out that $f: H \rightarrow K$ is a mapping of H onto K iff K is the range of f. Note that every "onto" mapping is an "into" mapping, but the converse is not necessarily true.

If $f:A \to B$, we know from Definition 7.6 that each point x of A maps onto exactly one point of B, namely the point $f(x)$ which is the image of x under the mapping f. It need not be the case, however, that different points of A map onto different points of B, as noted in Example 7.8. Mappings which do have the desirable property of mapping distinct points onto distinct points are called one-to-one mappings.

Definition 7.12

A mapping $f:H \to K$ is called *one-to-one* iff given any pair of points x, y of H such that $x \neq y$, we have $f(x) \neq f(y)$.

We shall have many occasions in the sequel to consider mappings which are both one-to-one and onto, and for convenience will refer to them as *bijections*. Specifically, a one-to-one mapping $f:A \xrightarrow{\text{onto}} B$ will be called a *bijection on A to B*.

Our next two theorems involve bijections. For the purpose of these two theorems only, we use the symbol H to denote the set of all nonnegative integers. Recalling that I denotes the set of positive integers and Z the set of all integers, we thus have

$$I = \{1,2,3, \ . \ . \ .\}$$
$$H = \{0,1,2,3, \ . \ . \ .\}$$
$$Z = \{. \ . \ . \ ,-3,-2,-1,0,1,2,3, \ . \ . \ .\}$$

Theorem 7.13

There is a bijection f on I to H.

Proof We define $f:I \to H$ by $f(n) = n - 1$ for each $n \in I$ and leave as an exercise the proof that f is a bijection on I to H. □

Theorem 7.14

There is a bijection g on H to Z.

Before proving this theorem, note that it follows intuitively quite simply, since we need only line up the nonnegative integers, place all the integers beneath them in the following way

0	1	2	3	4	5	6	· · ·
0	−1	1	−2	2	−3	3	· · ·

and then map each nonnegative integer onto the integer directly beneath it. The formal proof makes this process precise.

Proof The mapping g is defined by

$$g(n) = \begin{cases} \dfrac{n}{2} & \text{if } n \text{ is even and } n \geqslant 0 \\[2mm] -\dfrac{n+1}{2} & \text{if } n \text{ is odd and } n > 0 \end{cases}$$

The verification that g is a bijection on H to Z is left as an exercise. □

Example 7.15

Let $H = K = \{1,2,3,4,5\}$, and define f as follows:

$$f = \{\langle 1,1 \rangle, \langle 2,2 \rangle, \langle 3,3 \rangle, \langle 4,4 \rangle, \langle 5,5 \rangle\}$$

Here the domain and range of f are the same set, and the image under f of each element is just the element itself. This characterizes the identity mapping.

Definition 7.16

If f is the mapping of a set H onto itself defined by $f(x) = x$ for each $x \in H$, then f is called the *identity mapping on H*. The identity mapping of a set H onto itself is sometimes denoted by i_H.

Since mappings have been defined as sets, equality of mappings is merely a set equality. However, the following characterization will prove useful.

Theorem 7.17

If f and g are mappings, then $f = g$ iff f and g have the same domain A, and $f(x) = g(x)$ for every $x \in A$.

Proof Let $f: A \to B$ and $g: C \to D$, and denote their images by $f(A) = H$, $g(C) = K$. Then, $f \subset A \times H \subset A \times B$, where for each $x \in A$ there is exactly one $y \in H$ such that $\langle x,y \rangle \in f$. Similarly, $g \subset C \times K \subset C \times D$, where for each $x' \in C$ there is exactly one $y' \in K$ such that $\langle x',y' \rangle \in g$.

Now suppose that $f = g$. This means that $f \subset g$ and $g \subset f$. Let x be any point of A. Then there is exactly one $y \in H$ such that $\langle x,y \rangle \in f$. Since $f \subset g$, we see that $\langle x,y \rangle \in g$. It follows that $x \in C$ and $g(x) = y = f(x)$. Since x was an arbitrary element of A, we thus have $A \subset C$ and $g(x) = f(x)$ for every $x \in A$. A similar argument using $g \subset f$ yields $C \subset A$ and $f(x) = g(x)$ for every $x \in C$. Therefore $A = C$ and $f(x) = g(x)$ for every $x \in A$.

The converse is immediate, since if $A = C$ and $f(x) = g(x)$ for every $x \in A$, then $\langle x,y \rangle \in f$ iff $x \in A$ and $y = f(x)$ iff $x \in C$ and $y = g(x)$ iff $\langle x,y \rangle \in g$.   □

Suppose that we have a mapping $f: R_1 \to R_1$ defined by $f(x) = x^2$ for each $x \in R_1$ and a mapping $g: R_1 \to R_1$ defined by $g(x) = 3x$ for each $x \in R_1$. Let us look at a few real numbers and see what happens if we take their images under f and g. We see at once that $f(1) = 1$, $g(1) = 3$, $f(2) = 4$, $g(4) = 12$, $f(3) = 9$, $g(9) = 27$, $g(-4) = -12$, $f(-12) = 144$, and so on.

Now suppose we take a real number, say x, take its image under f, call it y, and then take the image of y under g. We can think of this process as finding the image of x under what we loosely call "the mapping f followed by the mapping g." Table 7.1 gives some examples.

Table 7.1

x	$y = f(x)$	$g(y)$	Image of x under "f followed by g"
1	1	3	3
-6	36	108	108
4	16	48	48
x	x^2	$3x^2$	$3x^2$
t	$f(t)$	$g[f(t)]$	$3t^2 = g[f(t)]$

Now let us make a similar table for the image of x under "the mapping g followed by the mapping f."

Table 7.2

x	$z = g(x)$	$f(z)$	Image of x under "g followed by f"
1	3	9	9
-6	-18	324	324
4	12	144	144
x	$3x$	$9x^2$	$9x^2$
t	$g(t)$	$f[g(t)]$	$9t^2 = f[g(t)]$

Two facts should be apparent from the above tables: (1) For each x, we can find the image of x under "f followed by g" by first finding $f(x)$ and then finding the image of $f(x)$ under g, which we quite naturally call $g[f(x)]$. Consequently, each x has exactly one image. (2) The image of x under "f followed by g" need not be the same as the image of x under "g followed by f." Clearly, for each x, we denote the image of x under "g followed by f" by $f[g(x)]$.

It is evident that in dealing with "f followed by g" we are working with a mapping, and we denote this mapping by $g \circ f$. Clearly, for each x, $(g \circ f)(x) = g[f(x)]$. In this example, we have worked with mappings f and g, each of which has R_1 as its domain. Suppose now that f has domain A and g has domain B, where A and B are proper subsets of R_1; that is,

$$f : A \longrightarrow R_1 \qquad \text{and} \qquad g : B \longrightarrow R_1$$

In this case, the mapping $g \circ f$ is defined only for those points x in A for which $f(x) \in B$. In particular, if the range of f is contained in the domain of g, then the domain of $g \circ f$ is A. As a general rule we have the following set inclusions:

$$\text{(domain of } g \circ f) \subset \text{(domain of } f)$$

$$\text{(range of } g \circ f) \subset \text{(range of } g)$$

These ideas are now formalized in the definition of a composite mapping.

Definition 7.18

Let A, B, C be nonempty sets and f, g mappings such that $f : A \to B$ and $g : B \to C$. Then the *composite mapping* of g with f is a mapping $h : A \to C$, where h is defined by $h(x) = g[f(x)]$ for every $x \in A$, and we write $h = g \circ f$.

The definition of composition can easily be extended to three or more mappings. For example, if $f : A \to B$, $g : B \to C$, and $h : C \to D$ are mappings, then we define the composite mapping $F : A \to D$, where $F = h \circ g \circ f$, as follows:

$$F(x) = h[g(f(x))] \qquad \text{for every } x \text{ in the domain } A \text{ of } f$$

Note that the expression within the brackets is the composite mapping of g with f, so that F is the composite mapping of h with $g \circ f$, which we may write as $h \circ (g \circ f)$. We remark that we shall get the same mapping F if we take the composite mapping of $h \circ g$ with f, written $(h \circ g) \circ f$. In asserting that $h \circ (g \circ f) = (h \circ g) \circ f$, we are really saying that composition of mappings is associative, and require proof of this fact in Prob. 7.6.

Example 7.19

Let $A = \{a,b,c,d,e,x,y,z\}$; $B = \{i,j,k,l,m\}$; $C = \{1,3,5,7\}$. Define the mapping $f : A \xrightarrow{\text{onto}} B$ as follows:

$$f(a) = k \qquad f(b) = m \qquad f(c) = i \qquad f(d) = j$$
$$f(e) = l \qquad f(x) = j \qquad f(y) = j \qquad f(z) = j$$

Now define the mapping $g : B \xrightarrow{\text{onto}} C$ as follows:

$$g(i) = 3 \qquad g(j) = 5 \qquad g(k) = 7$$
$$g(l) = 3 \qquad g(m) = 1$$

Let h be the composite mapping $g \circ f$. Then we obtain

$$h(a) = (g \circ f)(a) = g[f(a)] = g(k) = 7$$
$$h(b) = (g \circ f)(b) = g[f(b)] = g(m) = 1$$
$$h(c) = (g \circ f)(c) = g[f(c)] = g(i) = 3$$
$$h(d) = (g \circ f)(d) = g[f(d)] = g(j) = 5$$

The reader may verify in a similar fashion that $h(e) = 3$ and $h(x) = h(y) = h(z) = 5$. Thus we see that the composite mapping h is a mapping from A onto C; that is, $h = (g \circ f) : A \xrightarrow{\text{onto}} C$.

Example 7.20

Let $A, B, C, f, g,$ and h be defined as in Example 7.19. In addition, let $D = \{r,s,t\}$, and define the mapping $p : C \xrightarrow{\text{onto}} D$ as follows:

$$p(1) = r \qquad p(3) = s \qquad p(5) = t \qquad p(7) = r$$

Then the composite mapping $F = p \circ g \circ f$ maps A onto D. One way to determine the mapping F is to use the result of Example 7.19 and write $F = p \circ (g \circ f) = p \circ h$. For example, $F(a) = (p \circ h)(a) = p[h(a)] = p(7) = r$. The reader may verify, similarly, that $F(b) = r$, $F(c) = s$, $F(d) = t$, $F(e) = s$, $F(x) = t$, $F(y) = t$, and $F(z) = t$. Another way to determine F is to write $F = (p \circ g) \circ f$ and first find $(p \circ g) : B \xrightarrow{\text{onto}} D$. We note that

$$(p \circ g)(i) = p[g(i)] = p(3) = s$$
$$(p \circ g)(j) = p[g(j)] = p(5) = t$$

Similarly, $(p \circ g)(k) = r$, $(p \circ g)(l) = s$, and $(p \circ g)(m) = r$. Finally,

$$F(a) = [(p \circ g) \circ f](a) = (p \circ g)[f(a)] = (p \circ g)(k) = r$$
$$F(b) = [(p \circ g) \circ f](b) = (p \circ g)[f(b)] = (p \circ g)(m) = r$$
$$F(c) = [(p \circ g) \circ f](c) = (p \circ g)[f(c)] = (p \circ g)(i) = s$$

The reader may compute the remaining values of F and verify that they agree with those found above.

Example 7.21

Recall that I is the set of all positive integers, and let $B = \{1,3,5,7,9,11, \ldots\}$. Define $f: I \xrightarrow{\text{onto}} B$ by $f(x) = 2x - 1$ for each $x \in I$. Let $C = \{2,4,6,8,10, \ldots\}$, and define $g: B \to C$ by $g(y) = 2y$ for each $y \in B$. Note that g is a mapping of B into C which is not an onto mapping. The composite mapping $h: I \to C$ is defined by $h = g \circ f$; that is, $h(x) = g[f(x)]$ for every $x \in I$. In this particular example, $h(x) = g(2x - 1) = 4x - 2$ for every $x \in I$. Thus h maps I onto the set $H = \{2,6,10,14, \ldots\}$. Note that h maps I into C, not onto C, since H is a proper subset of C.

In Example 7.19 the composite mapping is an onto mapping, whereas in Example 7.21 the composite mapping is not an onto mapping. The next theorem presents a case in which we can be sure that a composite mapping is onto.

Theorem 7.22

If $f: A \xrightarrow{\text{onto}} B$ and $g: B \xrightarrow{\text{onto}} C$, then $(g \circ f): A \xrightarrow{\text{onto}} C$.

Proof We know that $g \circ f$ is a composite mapping from A into C by Definition 7.18. Thus we need only prove that $g \circ f$ is an onto mapping. We do this by showing that each element of C is the image under $g \circ f$ of at least one element of A. Thus let p be any element of C. Since g is an onto mapping, there exists a point q in B such that $g(q) = p$. But f is also onto, and so there exists a point r in A such that $f(r) = q$. Therefore,

$$p = g(q) = g[f(r)] = (g \circ f)(r) \qquad \square$$

The proofs of our next theorem and its corollaries are left as exercises.

Theorem 7.23

If $f: A \to B$ and $g: B \to C$ are one-to-one mappings, then $(g \circ f): A \to C$ is a one-to-one mapping.

Corollary 7.24

If f is a bijection on A to B and g is a bijection on B to C, then $g \circ f$ is a bijection on A to C.

Corollary 7.25

There is a bijection on I to Z.

Let us return for a moment to Examples 7.19 and 7.21. In Example 7.19, the range of g is not contained in the domain of f, and so we cannot define a mapping $f \circ g$. However, in Example 7.21, the range of g (the set of all positive even integers) is contained in the domain of f (the set of all positive integers), and so we can define the composite mapping $f \circ g$. Writing $B = \{2k - 1 \mid k \in I\}$, we have for any element in B

$$(f \circ g)(2k - 1) = f[g(2k - 1)] = f[2(2k - 1)] = f(4k - 2)$$
$$= 2(4k - 2) - 1 = 8k - 5$$

Thus the range of $f \circ g$ is the set $\{3,11,19,27, \ldots\}$, which is a proper subset of B.

We close this section by introducing a useful type of mapping called a retraction.

Definition 7.26

Let A and B be nonempty sets. The mapping $f : A \xrightarrow{\text{onto}} B$ is called a *retraction of A onto B* iff the following two conditions hold:

(a) B is a subset of A.
(b) $f(x) = x$ for every $x \in B$.

The set B is said to be a *retract* of A iff there is a retraction f of A onto B.

Theorem 7.27

Let A be any set, and let B be any nonempty subset of A. Then B is a retract of A.

Proof Using the fact that B is nonempty, let b be any point of B. We define a mapping f of A onto B as follows. For any $x \in A$,

$$f(x) = \begin{cases} x & \text{if } x \in B \\ b & \text{if } x \notin B \end{cases}$$

Clearly, f is a retraction of A onto B. $\square$

Example 7.28

Let $A = \{a,b,c,d\}$ and $B = \{a,c\}$. We can define a retraction $f : A \xrightarrow{\text{onto}} B$ as follows:

$$f(a) = a \qquad f(c) = c$$
$$f(b) = a \qquad f(d) = a$$

Of course, we could just as well have mapped each of the points b and d onto the point c.

PROBLEMS

7.1 Let $H = \{1,2,3,4\}$ and $K = \{1,2,3,4,5\}$.

(a) Display the elements of the Cartesian product $H \times K$.

(b) If $f : H \to K$ is defined by $f(x) = x + 1$ for each $x \in H$, what are the domain and range of f?

(c) Show that f is a subset of $H \times K$ by displaying the elements of f, and convince yourself that f satisfies condition (a) of Definition 7.6.

(d) Is f an onto mapping?

(e) Is f one-to-one?

7.2 (a) If $f : H \xrightarrow{\text{onto}} K$ is defined by $f(x) = x^3 + 1$, and if $K = \{2,9,28,65\}$, determine H, where H is a subset of the real numbers.

(b) Display the elements of the Cartesian product $H \times K$, and show that f is a subset of $H \times K$ which satisfies condition (a) of Definition 7.6.

(c) Is f one-to-one?

7.3 Repeat Prob. 7.2 if $f : H \xrightarrow{\text{onto}} K$ is defined by $f(x) = x^2 + 1$ and $K = \{2,5,10,17\}$.

7.4 Let $H = \{1,2,3\}$ and $K = \{a,b,c\}$. Display the Cartesian product $H \times K$ and the Cartesian product $K \times H$. Let f and g be mappings of H into K defined by

$$f = \{\langle 1,b \rangle, \langle 2,a \rangle, \langle 3,c \rangle\}$$
$$g = \{\langle 1,c \rangle, \langle 2,b \rangle, \langle 3,c \rangle\}$$

Let F and G be subsets of $K \times H$ defined by

$$F = \{\langle b,1 \rangle, \langle a,2 \rangle, \langle c,3 \rangle\}$$

$$G = \{\langle c,1 \rangle, \langle b,2 \rangle, \langle c,3 \rangle\}$$

(a) Explain how each of the sets F and G is obtained from the corresponding mapping f and g.

(b) Is either of these sets F and G a mapping? If so, what is its domain, and what is its range? If not, why not?

7.5 Let A and B be sets and f a mapping from A into B. Define F as the subset of $B \times A$ obtained from f as in Prob. 7.4. Can you guess a condition which will guarantee that F is a mapping from B into A?

7.6 Prove that the composition of three mappings f, g, h is associative.

7.7 Prove that the mapping f as defined in the proof of Theorem 7.13 is a bijection.

7.8 (a) Prove that the mapping g as defined in the proof of Theorem 7.14 is a bijection.

(b) Compute $g(20)$, $g(55)$, $g(243)$, and $g(1,000)$.

(c) Determine the nonnegative integer n such that $g(n) = -49$; $g(n) = 87$.

7.9 Prove Theorem 7.23.

7.10 Prove Corollary 7.24.

7.11 Prove Corollary 7.25.

7.12 According to Corollary 7.25, there is a bijection on I to Z, where Z is the set of all integers. Show that if we choose $F = g \circ f$, where f and g are the mappings defined in the proofs of Theorems 7.13 and 7.14, respectively, then F has the following definition:

$$F(n) = \begin{cases} \dfrac{n-1}{2} & \text{if } n \in I \text{ and } n \text{ is odd} \\[2mm] -\dfrac{n}{2} & \text{if } n \in I \text{ and } n \text{ is even} \end{cases}$$

Compute $F(7)$, $F(20)$, and $F(283)$. Determine $n \in I$ such that $F(n) = 40$, $F(n) = -83$, and $F(n) = 0$.

7.13 Show that if B is a retract of A and C is a retract of B, then C is a retract of A.

7.14 Let E be the set of all positive even integers. Is E a retract of I? If so, define a mapping $f : I \xrightarrow{\text{onto}} E$ which is a retraction.

7.15 Given $f : B \rightarrow H$ a one-to-one mapping and A a proper subset of B, prove that $f(A)$ is a proper subset of $\{f\} = f(B)$.

7.16 Prove that a mapping $f : A \rightarrow B$ is a constant mapping iff $\{f\}$ is a singleton.

7.17 Determine which of the following mappings f are one-to-one. Unless otherwise indicated, the domain of each f is the set of all real numbers x for which $f(x)$ is a real number. Determine the domain and range of each f.

(a) $\qquad\qquad\qquad f(x) = x + 1$

(b) $\qquad\qquad\qquad f(x) = x^2 + 1$

(c) $\qquad\qquad\qquad f(x) = x^3 + 1$

(d) $\qquad\qquad\qquad f(x) = \dfrac{1}{x}$

(e) $\qquad\qquad\qquad f(x) = \tan x$

(f) $\qquad\qquad\qquad f(x) = \tan x \qquad \text{for } 0 < \left| x - \dfrac{\pi}{2} \right| < \dfrac{\pi}{2}$

(g) $\qquad\qquad\qquad f(x) = \cos x \qquad \text{for } 0 \leqslant x \leqslant \pi$

(h) $\qquad\qquad\qquad f(x) = \sqrt{1 - x^2}$

8
Polynomials, Rational Functions, and Difference Quotients

We have seen that under appropriate conditions mappings can be combined under composition to yield new mappings. It is also possible to construct new mappings as algebraic combinations of given mappings.

Specifically, we use the algebraic operations of R_1 to define corresponding algebraic operations on the set of all mappings with range in R_1. It is an easy exercise to verify that the objects so obtained are mappings with range in R_1.

Definition 8.1

Let H be a subset of R_1, and suppose that $f : H \to R_1$ and $g : H \to R_1$ are mappings. We define the *sum* $f + g$, the *difference* $f - g$, the *product* fg, and the *quotient* f/g of the mappings f and g as follows. For each $x \in H$

(a) $(f + g)(x) = f(x) + g(x)$
(b) $(f - g)(x) = f(x) - g(x)$

70

(c) $(fg)(x) = f(x)g(x)$

(d) $(f/g)(x) = f(x)/g(x)$ provided $g(x) \neq 0$. If $g(c) = 0$ for some $c \in H$, the function f/g is not defined at c.

As a particular case of (c) above, suppose f is a constant mapping, say $f(x) = k$ for every $x \in H$. Then the product fg is given by

$$(fg)(x) = f(x)g(x) = kg(x)$$

for every $x \in H$. Thus the product of a constant and a mapping may be considered as a product of two mappings, one of which is a constant mapping. Note also that sums and products may be defined for more than two mappings using mathematical induction. We now illustrate some of these mappings.

Example 8.2

Let $H = \{x \in R_1 \mid x > 0\}$, and define f and g as follows: For each $x \in H$, $f(x) = x^2$ and $g(x) = x$. Then for each $x \in H$ we have $(f + g)(x) = x^2 + x$, $(f - g)(x) = x^2 - x$, $(fg)(x) = x^3$, and $(f/g)(x) = x$.

All the mappings of this example belong to a special class of functions called polynomials.

Definition 8.3

A *polynomial on a subset H of* R_1 is a function $f : H \to R_1$ defined by

$$f(x) = a_0 x^n + a_1 x^{n-1} + a_2 x^{n-2} + \cdots + a_{n-1} x + a_n$$

for every $x \in H$, where n is a nonnegative integer and $a_0, a_1, \ldots, a_n$ are real numbers with $a_0 \neq 0$. The integer n is called the *degree* of the polynomial, and for each integer i such that $0 \leq i \leq n$, $a_i x^{n-i}$ is called the *term of degree* $n - i$ of the polynomial and a_i is referred to as the *coefficient* of this term. When no domain is specified for a polynomial, we may assume that the domain is R_1.

Note that a constant mapping is a polynomial of degree zero. In Example 8.2, the functions g and f/g are polynomials of degree 1; f, $f + g$, and $f - g$ are polynomials of degree 2; and fg is a polynomial of degree 3. In general, the sum, difference, and product of two polynomials are again polynomials. It is easy to see that if f and g are polynomials of degrees m and n, respectively, then the degree of fg is $m + n$, and the degrees of $f + g$ and $f - g$ are $\leq \sup \{m, n\}$, with equality holding in these cases except when $m = n$ and the highest-powered terms of f and g cancel each other. The quotient f/g of the polynomials f and g is called a *rational function* and is generally not a polynomial. However, if there exists a polynomial h such that $f = gh$, then it is evident that $m \geq n$ and f/g is a polynomial of degree $m - n$. In this case we say that g is a *factor* of f. Clearly, h is also a factor of f and has degree $m - n$. We remark that even though $f = gh$, the polynomials f/g and h are not necessarily equal, since their domains may not be equal. In particular, any real number c for which $g(c) = 0$ is not in the domain of f/g. Consider the following example.

Example 8.4

Let f, g, h be polynomials defined on R_1 by

$$f(x) = x^2 - 1 \qquad g(x) = x - 1 \qquad h(x) = x + 1$$

It is evident that $f = gh$, so that g and h are factors of f. Note that f/g is not defined at the point $x = 1$, so that f/g and h are not equal. However, for every $x \in R_1 - \{1\}$, we have $(f/g)(x) = h(x)$. Thus the graph of the function f/g consists of the graph of the function h with the point $\langle 1,2 \rangle$ deleted. Similarly, the graph of the function f/h consists of the graph of g with the point $\langle -1,-2 \rangle$ deleted.

Our next theorem and corollary may be recognized as the factor and remainder theorems, respectively, of college algebra.

Theorem 8.5

If f is a polynomial and p is a real number, then $f(x) - f(p)$ has $x - p$ as a factor; that is, $f(x) - f(p) = (x - p) \cdot P(x)$, where P is a polynomial. Thus $x - p$ is a factor of $f(x)$ iff $f(p) = 0$.

Proof Let f be any polynomial. Then we may write

$$f(x) = a_0 x^n + a_1 x^{n-1} + \cdots + a_{n-1} x + a_n$$

Hence

$$f(p) = a_0 p^n + a_1 p^{n-1} + \cdots + a_{n-1} p + a_n$$

so that

$$f(x) - f(p) = a_0(x^n - p^n) + a_1(x^{n-1} - p^{n-1}) + \cdots + a_{n-1}(x - p)$$

By Corollary 2.17, $x - p$ is a factor of each term on the right side of the above equation and hence is a factor of the left side. Therefore $f(x) - f(p) = (x - p) \cdot P(x)$, where P is a polynomial of degree $n - 1$. $\quad \square$

Corollary 8.6

If f is a polynomial and p is a real number, then
$$f(x) = (x - p)P(x) + f(p)$$
that is, if $f(x)$ is divided by $x - p$, the remainder is $f(p)$.

It follows from Theorem 8.5 that if f is a polynomial and p is a real number, then there exists a polynomial P such that
$$\frac{f(x) - f(p)}{x - p} = P(x) \qquad \text{for } x \neq p$$

There is a special name and symbol for the expression on the left side of the above equation.

Definition 8.7

Let f be any function with domain A, and let p be any point of A. We shall use the symbol $Q_{f,p}$ to denote the function defined by
$$Q_{f,p}(x) = \frac{f(x) - f(p)}{x - p} \qquad \text{for } x \in A - \{p\}$$

$Q_{f,p}$ is called the *difference quotient of the function f at the point p*.

By virtue of this definition, the following theorem is an immediate consequence of Theorem 8.5.

Theorem 8.8

If f is a polynomial of degree n with domain (a,b), then for any $p \in (a,b)$ the difference quotient $Q_{f,p}$ is a polynomial of degree $n - 1$ with domain (a,p,b).

Definition 8.9

A mapping $f : A \to R_1$ is said to be *bounded* iff its range $\{f\}$ is a bounded subset of R_1.

Our next result follows immediately from Theorem 3.13.

Theorem 8.10

A mapping $f : A \to R_1$ is bounded iff there exists a positive number M such that $\{f\} \subset (-M, M)$.

The positive number M of Theorem 8.10 is called a *bound* for the mapping f on the set A.

The following two lemmas present results which are easy consequences of Definition 8.9 and our work with bounded sets in Chap. 3. Proofs of these results are requested in the problems.

Lemma 8.11

Let $f:A \to R_1$ and $g:A \to R_1$ be bounded mappings, and let k be a real number. Then each of the following mappings is bounded:

(a) kf
(b) $f+g$
(c) $f-g$
(d) fg
(e) The mapping $h:A \to R_1$ defined by $h(x) = f(x) + k$ for each $x \in A$

Lemma 8.12

If A is a bounded subset of R_1, then the identity mapping $i_A :A \to R_1$ is bounded.

Theorem 8.13

Let A be a set and for any positive integer n, let $a_1, a_2, \ldots, a_n$ be real numbers, and let $f_1, f_2, \ldots, f_n$ be a collection of bounded mappings of A into R_1. Then each of the following mappings of A into R_1 is bounded:

(a) $$f_1 + f_2 + \cdots + f_n$$
(b) $$a_1 f_1 + a_2 f_2 + \cdots + a_n f_n$$
(c) $$a_1 f_1 f_2 \cdots f_n$$

Proof (a) follows easily from Lemma 8.11b using mathematical induction. (b) is then immediate from (a) and Lemma 8.11a. Part (c) follows from parts (a) and (d) of Lemma 8.11 using mathematical induction.   □

Corollary 8.14

Let A be a bounded subset of R_1 and n a positive integer. Then each of the following is a bounded mapping:

(a) The mapping $f:A \to R_1$ defined by $f(x) = x$ for $x \in A$
(b) The mapping $g:A \to R_1$ defined by $g(x) = x^n$ for $x \in A$
(c) Every polynomial $f:A \to R_1$

Corollary 8.15

Let f be a polynomial defined on a bounded set A. Then there exists a positive real number M such that $|f(x)| \leqslant M$ for every $x \in A$.

It is useful to be able to find a specific value for the bound M whose existence is guaranteed by Corollary 8.15. Although the value of M depends on the polynomial f and its domain A, the simple procedure for determining M is the same for all polynomials with bounded domains. An example will illustrate this procedure.

Example 8.16

If $A = (-3,5)$ and $f:A \to R_1$ is defined by

$$f(x) = 2x^3 - 3x^2 + 4x - 12$$

find a real number M such that $|f(x)| \leqslant M$ for every $x \in A$.

Solution Clearly, $|x| < 5$ for every $x \in A$. Thus for any $x \in A$ we have

$$\begin{aligned}
|f(x)| &\leqslant 2|x^3| + 3|x^2| + 4|x| + |-12| \\
&< 2(5)^3 + 3(5)^2 + 4(5) + 12 \\
&= 250 + 75 + 20 + 12 \\
&= 357
\end{aligned}$$

We may thus choose $M = 357$ or any larger number. It should be carefully noted, however, that we make no guarantee that 357 is the smallest possible value of M. In fact, we shall see in Prob. 8.7 that 305 is a bound for f on A.

Theorem 8.17

Let f be a polynomial defined on a bounded set A, and let p be a point of A. Then there exists a positive real number M such that

(a) $|f(x) - f(p)| \leqslant M|x - p|$ for every $x \in A$

and

(b) $|Q_{f,p}(x)| \leqslant M$ for every $x \in A - \{p\}$

Proof By virtue of Theorem 8.5 we have for each $x \in A$

$$|f(x) - f(p)| = |x - p| \cdot |P(x)|$$

where P is a polynomial on the bounded set A. Hence by Corollary 8.15, there exists a positive real number M such that $|P(x)| \leqslant M$ for every $x \in A$. This proves (a). For (b), we need merely note that $Q_{f,p}(x) = P(x)$ for $x \in A - \{p\}$. $\square$

Example 8.18

If $A = (-3,5)$ and $f:A \rightarrow R_1$ is defined by

$$f(x) = 4x^3 - 7x^2 + 19x - 7$$

find a positive real number M such that

$$|f(x) - f(4)| \leqslant M|x - 4| \qquad \text{for every } x \in A$$

Solution Since $f(x) = 4x^3 - 7x^2 + 19x - 7$, we have

$$f(4) = 4(4)^3 - 7(4)^2 + 19(4) - 7$$

and hence

$$\begin{aligned} f(x) - f(4) &= 4(x^3 - 4^3) - 7(x^2 - 4^2) + 19(x - 4) \\ &= (x - 4)[4(x^2 + 4x + 16) - 7(x + 4) + 19] \\ &= (x - 4)(4x^2 + 9x + 55) \end{aligned}$$

Using the technique developed in Example 8.16, it is easy to see that for every $x \in A$

$$|4x^2 + 9x + 55| < 200$$

Therefore,

$$|f(x) - f(4)| \leqslant 200|x - 4| \qquad \text{for every } x \in A$$

PROBLEMS

8.1 Let f, g, h be polynomials defined on R_1 by

$$f(x) = x^3 + 1 \qquad g(x) = x^2 + 1 \qquad h(x) = x + 1$$

(a) Determine $f + g$, $f - h$, gh, and $f \circ g$.
(b) Is f/g a polynomial? What is its domain?
(c) Is f/h a polynomial? What is its domain?
(d) Determine $Q_{f,p}(x)$ for $x \in R_1 - \{p\}$ and $Q_{h,3}(x)$ for $x \in R_1 - \{3\}$.

8.2 Prove Lemma 8.11.

8.3 Prove Lemma 8.12.

8.4 Carry out the mathematical induction in the proofs of (a) and (c) in Theorem 8.13.

8.5 Show that $x - 2$ is a factor of the polynomial $2x^3 - 3x^2 + 4x - 12$, and verify that $2x^3 - 3x^2 + 4x - 12 = (x - 2)(2x^2 + x + 6)$.

8.6 Show that if $-3 < x < 5$, then $|x - 2| < 5$.

8.7 If $A = (-3,5)$ and $f:A \rightarrow R_1$ is defined by $f(x) = 2x^3 - 3x^2 + 4x - 12$, use Probs. 8.5 and 8.6 to show that 305 is a bound for f on A.

8.8 Using A and f as defined in Prob. 8.7, find a positive real number M such that $|f(x) - f(3)| \leqslant M|x - 3|$.

8.9 For a fixed real number p, if we define the real number h by $h = x - p$ for $x \in R_1$, verify that $h < 0$ iff $x < p$; $h > 0$ iff $x > p$; and $h = 0$ iff $x = p$.

8.10 Using h as defined in Prob. 8.9 show that the difference quotient $Q_{f,p}$ can be expressed as a function of h in the form

$$Q_{f,p}(h) = \frac{f(p+h) - f(p)}{h}$$

8.11 Using the inequality derived in Example 8.18, show that:
  (a) If $|x - 4| < \frac{1}{200}$, then $|f(x) - f(4)| < 1$.
  (b) If $|x - 4| < \frac{1}{400}$, then $|f(x) - f(4)| < \frac{1}{2}$.
  (c) If $\epsilon > 0$ and $|x - 4| < \epsilon/200$, then $|f(x) - f(4)| < \epsilon$.

8.12 If f and g are polynomials of degrees m and n, respectively, what can be said about the composition $f \circ g$?

8.13 Are there any polynomials f for which $f \circ f = ff$?

9
Continuity

We begin this chapter by reviewing Example 8.18, where we found that if $A = (-3,5)$ and if f is defined by $f(x) = 4x^3 - 7x^2 + 19x - 7$, then

$$|f(x) - f(4)| \leq 200|x - 4| \qquad \text{for every } x \in A$$

This inequality exhibits the behavior of the function f near the point $p = 4$ in its domain. In fact, it is evident that if we are given any real number $\epsilon > 0$, then $|f(x) - f(4)| < \epsilon$ for every $x \in A$ satisfying $|x - 4| < \epsilon/200$. In other words, we can get $f(x)$ as "close as desired" to $f(4)$, that is, within any preassigned positive distance ϵ, merely by choosing x "sufficiently close" to 4, that is, within the positive distance $\delta = \epsilon/200$. We express this well-behaved property of the function by asserting that f is continuous at the point $p = 4$.

Definition 9.1

Let f be a function with domain A, and let p be a point of A. Then f is said to be *continuous at the point* p iff given any real number $\epsilon > 0$, there exists a real number $\delta > 0$ such that

$$|f(x) - f(p)| < \epsilon \qquad \text{for every } x \in A \text{ satisfying } |x - p| < \delta$$

f is said to be *continuous* (or *continuous on* A) iff f is continuous at p for every $p \in A$. A function which is not continuous is said to be *discontinuous*, and a point p at which f is discontinuous is called a *discontinuity* (or a *point of discontinuity*) of f.

Theorem 9.2

Every polynomial is continuous.

Proof Let f be a polynomial with domain A, and let p be any point of A. Since we want to show that f is continuous at p, let $\epsilon > 0$ be given. Define the set $B = A \cap (p - 1, p + 1)$, noting that B is bounded and $p \in B$. By Theorem 8.17a, there exists a positive real number M such that

$$|f(x) - f(p)| \leq M|x - p| \qquad \text{for every } x \in B$$

Thus $|f(x) - f(p)| < \epsilon$ for every $x \in B$ satisfying $|x - p| < \epsilon/M$. However, we know that $x \in B$ iff $x \in A$ and $|x - p| < 1$. Choosing $\delta = \inf \{1, \epsilon/M\}$, it follows that

$$|f(x) - f(p)| < \epsilon \quad \text{for every } x \in A \text{ satisfying } |x - p| < \delta \qquad \square$$

Lemma 9.3

If a function f with domain A is continuous at a point $p \in A$, then given any real number $\epsilon > 0$, there exists a real number $\delta > 0$ such that

$$|f(p)| - \epsilon < |f(x)| < |f(p)| + \epsilon \qquad \text{for every } x \in A \cap (p - \delta, p + \delta)$$

Proof Follows immediately from the definition of continuity and Theorem 3.12i. $\square$

Theorem 9.4

If a function f with domain A is continuous at a point $p \in A$, then there is an open interval containing p on which f is bounded; specifically, there exist real numbers $\delta > 0$ and $M > 0$ such that

$$|f(x)| \leq M \qquad \text{for every } x \in A \cap (p - \delta, p + \delta)$$

Proof Given $\epsilon > 0$, choose $M = |f(p)| + \epsilon$ and use Lemma 9.3. $\square$

Theorem 9.5

If $f : A \to R_1$ and $g : A \to R_1$ are continuous, then each of the following is a continuous function of A into R_1:

(a) $f + g$
(b) fg
(c) $f - g$

Proof   For (a), let p be any point of A, and let $\epsilon > 0$ be given. Since f is continuous at p, there exists a real number $\delta_1 > 0$ such that

$$|f(x) - f(p)| < \frac{\epsilon}{2} \qquad \text{for every } x \in A \text{ satisfying } |x - p| < \delta_1$$

and since g is continuous at p, there exists a real number $\delta_2 > 0$ such that

$$|g(x) - g(p)| < \frac{\epsilon}{2} \qquad \text{for every } x \in A \text{ satisfying } |x - p| < \delta_2$$

Defining $\delta = \inf \{\delta_1, \delta_2\}$, it follows that

$$\begin{aligned}
|(f + g)(x) - (f + g)(p)| &= |f(x) + g(x) - f(p) - g(p)| \\
&= |[f(x) - f(p)] + [g(x) - g(p)]| \\
&\leq |f(x) - f(p)| + |g(x) - g(p)| \\
&< \frac{\epsilon}{2} + \frac{\epsilon}{2} \\
&= \epsilon
\end{aligned}$$

for every $x \in A$ satisfying $|x - p| < \delta$. Thus $f + g$ is continuous at p, which proves (a).

For (b), let p be any point of A, and let $\epsilon > 0$ be given. We first choose M_1 such that $|g(p)| < M_1$. Since f is continuous at p, there exists a real number $\delta_1 > 0$ such that

$$|f(x) - f(p)| < \frac{\epsilon}{2M_1} \qquad \text{for every } x \in A \text{ satisfying } |x - p| < \delta_1$$

Defining $M_2 = |f(p)| + \epsilon/(2M_1)$ it follows from Lemma 9.3 that

$$|f(x)| < M_2 \qquad \text{for every } x \in A \text{ satisfying } |x - p| < \delta_1$$

Since g is continuous at p, there exists a real number $\delta_2 > 0$ such that

$$|g(x) - g(p)| < \frac{\epsilon}{2M_2} \qquad \text{for every } x \in A \text{ satisfying } |x - p| < \delta_2$$

Defining $\delta = \inf \{\delta_1, \delta_2\}$, we thus have

$$\begin{aligned}
|(fg)(x) - (fg)(p)| &= |f(x)g(x) - f(p)g(p)| \\
&= |f(x)[g(x) - g(p)] + g(p)[f(x) - f(p)]| \\
&\leq |f(x)||g(x) - g(p)| + |g(p)||f(x) - f(p)| \\
&< M_2 \cdot \frac{\epsilon}{2M_2} + M_1 \cdot \frac{\epsilon}{2M_1} \\
&= \epsilon
\end{aligned}$$

for every $x \in A$ satisfying $|x - p| < \delta$. Therefore fg is continuous at p, which proves (b).

Note that (c) follows from (a) and (b) by writing

$$f - g = f + (-1)g \qquad \square$$

We state Theorem 9.5 in words by asserting that the sum, product, and difference of two continuous functions are again continuous functions. These results can easily be extended to more than two functions by mathematical induction.

Corollary 9.6

For any positive integer n, let $f_1, f_2, \ldots, f_n$ be a collection of continuous functions of A into R_1. Then each of the following is a continuous function of A into R_1:

(a) $f_1 + f_2 + \cdots + f_n$
(b) $f_1 f_2 \cdots f_n$

Our next theorem will show that the quotient of two continuous functions is continuous wherever the quotient is defined. Its proof will use the following lemma.

Lemma 9.7

If a function g with domain A is continuous at the point $p \in A$, and if $g(p) \neq 0$, then the function $1/g$ is continuous at p.

Proof Let $\epsilon > 0$ be given, and let $|g(p)| = k > 0$. By Lemma 9.3, there exists a real number $\delta_1 > 0$ such that

$$|g(p)| - \frac{k}{2} < |g(x)| < |g(p)| + \frac{k}{2}$$

for every $x \in A$ satisfying $|x - p| < \delta_1$

In particular, since $|g(p)| = k$, we see that $|g(x)| > k/2$ and hence

$$\frac{1}{|g(x)|} < \frac{2}{k}$$

for every $x \in A$ satisfying $|x - p| < \delta_1$. Now we again use the continuity of g at p to assert the existence of a real number $\delta_2 > 0$ such that

$$|g(x) - g(p)| < \frac{k^2 \epsilon}{2} \qquad \text{for every } x \in A \text{ satisfying } |x - p| < \delta_2$$

Defining $\delta = \inf \{\delta_1, \delta_2\}$, we thus have

$$\left|\frac{1}{g}(x) - \frac{1}{g}(p)\right| = \left|\frac{1}{g(x)} - \frac{1}{g(p)}\right|$$
$$= \left|\frac{g(p) - g(x)}{g(x)g(p)}\right|$$
$$= |g(x) - g(p)| \cdot \frac{1}{|g(x)|} \cdot \frac{1}{|g(p)|}$$
$$< \frac{k^2\epsilon}{2} \cdot \frac{2}{k} \cdot \frac{1}{k}$$
$$= \epsilon$$

for every $x \in A$ satisfying $|x - p| < \delta.$ $\quad\square$

Theorem 9.8

If $f : A \to R_1$ and $g : A \to R_1$ are continuous, then the quotient f/g is continuous at every point $x \in A$ for which $g(x) \neq 0$.

Proof Let p be any point in the domain of f/g, noting from the definition of a quotient that $g(p) \neq 0$. Then $1/g$ is continuous at p by Lemma 9.7, and hence f/g (which is the product of f and $1/g$) is continuous at p by Theorem 9.5b. $\square$

Corollary 9.9

Every rational function is continuous on its domain.

Corollary 9.10

If a function f is continuous on its domain A, then for any point $p \in A$ the difference quotient $Q_{f,p}$ is continuous on $A - \{p\}$.

Corollary 9.11

For any point $p \in R_1$, the mapping g_p defined by $g_p(x) = 1/(x - p)$, is continuous on its domain $M_p = R_1 - \{p\}$.

Theorem 9.12

The composition of two continuous functions is continuous; i.e., if $f : A \to B$ and $g : B \to C$ are continuous, then $(g \circ f) : A \to C$ is continuous.

Proof Let p be any point of A, and let $\epsilon > 0$ be given. Writing $q = f(p)$ and noting that g is continuous at the point q of B, we see that there exists a real number $\delta_1 > 0$ such that

$$|g(y) - g(q)| < \epsilon \qquad \text{for every } y \in B \text{ satisfying } |y - q| < \delta_1 \quad (1)$$

Also, since f is continuous at p, there exists a real number $\delta > 0$ such that

$$|f(x) - f(p)| < \delta_1 \quad \text{for every } x \in A \text{ satisfying } |x - p| < \delta \quad (2)$$

We have defined $f(p) = q$, and we know that for each $x \in A$, $f(x) = y$ for some $y \in B$. Thus for every $x \in A$ satisfying $|x - p| < \delta$, it follows from (2) that $|f(x) - q| < \delta_1$, and hence it follows from (1) that $|g[f(x)] - g(q)| < \epsilon$, or $|g[f(x)] - g[f(p)]| < \epsilon$. Therefore $|(g \circ f)(x) - (g \circ f)(p) < \epsilon$, and hence $g \circ f$ is continuous at p. $\quad \square$

Corollary 9.13

Let n be any positive integer, let $A_1, A_2, \ldots, A_{n+1}$ be nonempty subsets of R_1, and for each integer i such that $1 \leqslant i \leqslant n$ let f_i be a continuous function from A_i into A_{i+1}. Then the composition

$$f_n \circ f_{n-1} \circ \cdots \circ f_2 \circ f_1$$

is a continuous function from A_1 into A_{n+1}.

Theorem 9.14 Cauchy criterion for continuity of f at a point

A function $f : A \to R_1$ is continuous at the point $p \in A$ iff the following condition is satisfied:

(a) Given any real number $\epsilon > 0$, there exists a real number $\delta > 0$ such that $|f(x) - f(x')| < \epsilon$ for any points $x, x' \in A$ satisfying $|x - p| < \delta$ and $|x' - p| < \delta$.

Proof If condition (a) is satisfied, we need merely choose $x' = p$ to show that f is continuous at p. For the converse, suppose f is continuous at p, and let $\epsilon > 0$ be given. Then there exists a real number $\delta > 0$ such that

$$|f(x) - f(p)| < \frac{\epsilon}{2} \quad \text{for every } x \in A \text{ satisfying } |x - p| < \delta$$

Thus for any points $x, x' \in A$ satisfying $|x - p| < \delta$ and $|x' - p| < \delta$, we have

$$\begin{aligned} |f(x) - f(x')| &= |f(x) - f(p) + f(p) - f(x')| \\ &\leqslant |f(x) - f(p)| + |f(x') - f(p)| \\ &< \frac{\epsilon}{2} + \frac{\epsilon}{2} \\ &= \epsilon \quad \square \end{aligned}$$

It is evident from Definition 9.1 that a function f cannot be continuous at any point p which is not in the domain A of f. In fact, if $p \notin A$, the symbol $f(p)$ is meaningless. Thus f *must be defined at* p *in order for* f *to be continuous at* p. This condition, however, is far from sufficient for continuity at p, as we shall see below. We first characterize a discontinuity of a function f at a point p of its domain.

Theorem   9.15

Let f be a function with domain A, and let p be a point of A. Then f is discontinuous at p iff there exists a real number $\epsilon > 0$ such that given any $\delta > 0$, there is at least one point $x \in A$ satisfying

$$|x - p| < \delta \qquad \text{and} \qquad |f(x) - f(p)| \geq \epsilon$$

This theorem is merely the formal negation of Definition 9.1, and it is suggested that the interested reader convince himself of this fact. We now illustrate some discontinuities by means of examples.

Example   9.16

Let f be defined on R_1 by $f(1) = 1$ and $f(x) = x + 1$ for $x \in R_1 - \{1\}$. The graph of f is shown in Fig. 9.1. Clearly, on $R_1 - \{1\}$, f is a polynomial and hence is continuous. However, f is not continuous at the point $p = 1$, which we verify using Theorem 9.15 with $\epsilon = 1$. Note that for any $x > 1$, $f(x) > 2$. Thus given any $\delta > 0$, we need merely choose $x = 1 + \delta/2$, from which it follows that

$$|x - 1| < \delta$$

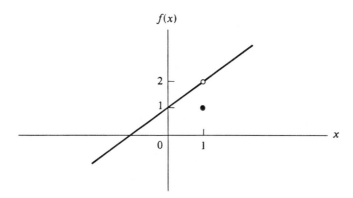

Figure 9.1

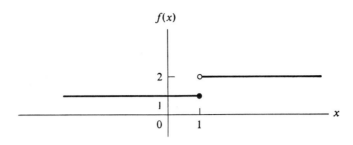

Figure 9.2

and

$$|f(x) - f(1)| = f\left(1 + \frac{\delta}{2}\right) - 1 > 2 - 1 = 1 = \epsilon$$

Example 9.17

Let f be defined on R_1 by

$$f(x) = \begin{cases} 1 & \text{for } x \leq 1 \\ 2 & \text{for } x > 1 \end{cases}$$

The graph of f is shown in Fig. 9.2. Again f is continuous on $R_1 - \{1\}$, and it is easy to show that f is discontinuous at the point $p = 1$, using $\epsilon = \frac{1}{2}$. For given $\delta > 0$ we again choose $x = 1 + \delta/2$, and hence

$$|x - 1| < \delta$$

and

$$|f(x) - f(1)| = |2 - 1| = 1 > \epsilon$$

Note that the same argument would work for any ϵ such that $0 < \epsilon < 1$.

Suppose now that we change the domain of f in Example 9.17 to the closed ray $A = \{x \in R_1 \mid x \leq 1\}$. Then f is a constant function and hence is continuous on A. In particular, f is continuous at the point $p = 1$. Of course, f is no longer defined for any points $x > 1$; therefore in applying the definition of continuity at the point $p = 1$ we consider only those points x of A satisfying $1 - \delta < x \leq 1$ or, equivalently, those x in A for which $0 \leq 1 - x < \delta$. This leads to the notion of one-sided continuity.

Definition 9.18

Let f be a function with domain A, and let p be a point of A. Then f is said to be:

(a) *Continuous from the left at p* iff given any real number $\epsilon > 0$ there exists a real number $\delta > 0$ such that

$$|f(x) - f(p)| < \epsilon \qquad \text{for every } x \in A \text{ satisfying } 0 \leqslant p - x < \delta$$

(b) *Continuous from the right at p* iff given any real number $\epsilon > 0$ there exists a real number $\delta > 0$ such that

$$|f(x) - f(p)| < \epsilon \qquad \text{for every } x \in A \text{ satisfying } 0 \leqslant x - p < \delta$$

The following theorem should now be self-evident.

Theorem 9.19

A function f is continuous at a point p iff f is both continuous from the left at p and continuous from the right at p.

We may now assert that the function f of Example 9.17 is continuous from the left but is not continuous from the right at the point $p = 1$. The function f of Example 9.16 is discontinuous from both right and left at $p = 1$. It should be clear that Theorem 9.4 has an obvious modification to one-sided continuity, merely by replacing the open interval $(p - \delta, p + \delta)$ by the appropriate half-open interval $(p - \delta, p]$ or $[p, p + \delta)$. Similarly, our other general results on continuity (particularly Theorems 9.5, 9.8, and 9.12) have obvious modifications to one-sided continuity.

PROBLEMS

9.1 Show that the function f defined by $f(x) = |x|$ is continuous on R_1.

9.2 Determine the difference quotient $Q_{f,0}$ for the function f of Prob. 9.1.

9.3 Defining the function $|f|$ by $|f|(x) = |f(x)|$ for every x in the domain A of a function f, show that if f is continuous on A, then $|f|$ is continuous on A. Is the converse true? If so, prove it. If not, give a counterexample.

9.4 Give examples to show that the sum, product, difference, and quotient of two functions may be continuous, even though the two given functions are not continuous.

9.5 Is it possible for $f \circ f$ to be continuous even though f is not continuous?

9.6 Show that for $p \in R_1$, the mapping g_p defined in Corollary 9.11 is a bijection on M_p to M_0.

9.7 Let a, p, b be real numbers such that $a < p < b$. Let f be a continuous function with domain $(a,p]$ and g a continuous function with domain $[p,b)$. Suppose that $f(p) = g(p)$. Define a function h with domain (a,b) by

$$h(x) = \begin{cases} f(x) & \text{if } x \in (a,p] \\ g(x) & \text{if } x \in [p,b) \end{cases}$$

Prove that h is continuous on (a,b).

10
A Bijection on I to Q; Inverse Mappings

In this chapter and the next we consider in detail some special mappings. These investigations serve the dual purpose of gaining experience in dealing with various mappings and developing some machinery that will be useful later in the text.

We begin with a slight digression to introduce the fundamental theorem of arithmetic. The proof of this theorem may be found in any text on elementary number theory and will not be given here. A simple illustration, however, may help to clarify the notation. First, we should recall that an integer $p > 1$ is a *prime* iff it has no divisor d, where d is an integer such that $1 < d < p$. The reader may easily verify that

$$17{,}640 = 8 \cdot 9 \cdot 5 \cdot 49 = 2^3 \cdot 3^2 \cdot 5^1 \cdot 7^2$$

Note that the factors 2, 3, 5, and 7 are primes and are raised to powers which are positive integers. We call this procedure the *factorization* of the integer 17,640 into a product of powers of primes. The fundamental theorem of arithmetic asserts that every integer $m > 1$ has such a factorization and furthermore that the factorization is unique.

Theorem 10.1 Fundamental theorem of arithmetic

Let m be an integer greater than 1. Then m can be written in exactly one way in the form

$$m = p_1{}^{n_1}\, p_2{}^{n_2} \cdots p_k{}^{n_k}$$

where $p_i < p_{i+1}$ for $i = 1, 2, \ldots, k - 1$, and for each $i = 1, 2, \ldots, k$, p_i is a prime and n_i is a positive integer.

We shall also make use of the inverse of a bijection. Recall that a one-to-one mapping $f : A \xrightarrow{\text{onto}} B$ is called a bijection on A to B. If f is such a mapping, then for each point q of B there is a unique point p of A such that $f(p) = q$. (There must be at least one such point since f is an onto mapping, and there must be at most one such point since f is one-to-one.)

Definition 10.2

Let f be a bijection on A to B, and for each point q of B denote by $f^{-1}(q)$ the unique point of A such that $f[f^{-1}(q)] = q$. Then the *inverse of the bijection f* is the mapping $g : B \to A$ defined by $g(q) = f^{-1}(q)$ for each point q of B. We denote by $f^{-1} : B \to A$ the inverse of the mapping f.

Although we have defined an inverse only for a mapping which is a bijection, we remark that every one-to-one mapping has an inverse. For if f is a one-to-one mapping of A into B, then $\{f\} = f(A)$ is a subset of B and f is a bijection on A to $\{f\}$. Clearly, f^{-1} exists, but its domain is $\{f\}$ rather than B. Thus the requirement $\{f\} = B$ is merely a notational convenience.

The proof of the following theorem is a simple exercise which follows at once from Definition 10.2.

Theorem 10.3

If f is a bijection on A to B, then:

(a) f^{-1} is a bijection on B to A.
(b) $f^{-1} \circ f = i_A$, the identity mapping on A (see Definition 7.16).
(c) $f \circ f^{-1} = i_B$.

A large class of mappings which always have inverses are the strictly monotone mappings included in our next definition.

Definition 10.4

Let A be a nonempty subset of R_1. A mapping $f:A \rightarrow R_1$ is said to be:

(a) *Nondecreasing* iff $f(x) \leqslant f(y)$
(b) *Increasing* iff $f(x) < f(y)$
(c) *Nonincreasing* iff $f(x) \geqslant f(y)$
(d) *Decreasing* iff $f(x) > f(y)$

whenever $x,y \in A$ and $x < y$

A mapping satisfying any one of these four conditions is called *monotone*, and a mapping satisfying (b) or (d) is called *strictly monotone*.

Note that a constant mapping is monotone but not strictly monotone. The mapping g_p of Corollary 9.11 is strictly monotone on each of the rays $\{x \mid x < p\}$ and $\{x \mid x > p\}$ but is *not* monotone on its entire domain M_p. Incidentally, this is a nice example of a mapping which is not strictly monotone and yet has an inverse. We consider another such mapping in Prob. 10.10. It is thus evident that strict monotonicity is a sufficient but not a necessary condition for the existence of an inverse.

Example 10.5

Let $H = \{x \in R_1 \mid x \geqslant 0\}$, and define $f:H \xrightarrow{\text{onto}} H$ by $f(x) = x^2$. It is easy to see that f is increasing on H, since if x_1 and x_2 are any two points of H such that $x_1 < x_2$, then

$$f(x_2) - f(x_1) = x_2{}^2 - x_1{}^2 = (x_2 - x_1)(x_2 + x_1) > 0$$

Hence f^{-1} exists. Since for any $x \in H$ we must have $f^{-1}[f(x)] = x$, it follows that $f^{-1}(x^2) = x$. But we know that for nonnegative x, $\sqrt{x^2} = x$, and so f^{-1} must be the nonnegative-square-root function, which is defined by $f^{-1}(x) = \sqrt{x}$ for $x \geqslant 0$.

Example 10.6

Let us determine the inverse of the bijection g on H to Z defined in the proof of Theorem 7.14. Recall that H represents the set of nonnegative integers and Z the set of all integers; g is defined by

$$g(n) = \begin{cases} \dfrac{n}{2} & \text{if } n \text{ is even and } n \geqslant 0 \\ -\dfrac{n+1}{2} & \text{if } n \text{ is odd and } n > 0 \end{cases}$$

Since we know from Theorem 10.3 that g^{-1} is a bijection on Z to H, we define g^{-1} by specifying its value for each element of Z. For each integer $q \in Z$, Definition 10.2 asserts that $g^{-1}(q)$ is the unique point in H such that $g[g^{-1}(q)] = q$. But according to the above definition of g, if q is nonnegative, then

$$g[g^{-1}(q)] = \frac{g^{-1}(q)}{2}$$

from which it follows that $g^{-1}(q) = 2q$; whereas if q is negative, then

$$g[g^{-1}(q)] = -\frac{g^{-1}(q) + 1}{2}$$

from which it follows that $g^{-1}(q) = -1 - 2q$. We thus have

$$g^{-1}(n) = \begin{cases} 2n & \text{if } n \geqslant 0 \\ -1 - 2n & \text{if } n < 0 \end{cases}$$

An interesting exercise is the verification that $g^{-1} \circ g = i_H$ and that $g \circ g^{-1} = i_Z$, as asserted in Theorem 10.3.

Example 10.7

It will also be convenient to have an explicit formula for the inverse of the bijection F on I to Z, where Z represents the set of all integers, defined in Prob. 7.12 by

$$F(n) = \begin{cases} \dfrac{n-1}{2} & \text{if } n \in I \text{ and } n \text{ is odd} \\ -\dfrac{n}{2} & \text{if } n \in I \text{ and } n \text{ is even} \end{cases}$$

We leave the details to the reader and merely state that the bijection F^{-1} on Z to I is defined by

$$F^{-1}(n) = \begin{cases} 2n + 1 & \text{if } n \geqslant 0 \\ -2n & \text{if } n < 0 \end{cases}$$

The interesting proof of our next theorem, due to Gerald Freilich, appeared in the November 1965 issue of *The American Mathematical Monthly*. Recall that a *rational number* is a real number which can be expressed as the quotient of two integers; i.e., the number r is rational iff $r = p/q$, where p and q are integers with $q \neq 0$.

Theorem 10.8

There is a bijection on I to the set Q of all rational numbers.

Proof We know that the mapping F of Example 10.7 is a bijection on I to the set Z of all integers. We shall now determine a bijection

h on Z to the set Q. It will then follow from Corollary 7.24 that the composition $k = h \circ F$ is a bijection on I to Q, thus completing the proof. We first define h on the set of nonnegative integers as follows:

$$h(0) = 0 \quad \text{and} \quad h(1) = 1$$

For any integer $m > 1$, we factor m as in Theorem 10.1 and define

$$h(m) = p_1{}^{g(n_1)} p_2{}^{g(n_2)} \cdots p_j{}^{g(n_j)}$$

where g is the mapping defined in Example 10.6. We next extend the definition of h to the negative integers. For any integer $m < 0$ we define

$$h(m) = -h(-m)$$

It is left as an exercise to show that h is a bijection on Z to Q. □

We illustrate the proof of Theorem 10.8 by determining the images of specific positive integers under the bijection k on I to Q. For example, suppose we want to find $k(27)$. Since $k(27) = h[F(27)] = h(13)$, and 13 is a prime number with factored form $13 = 13^1$, it follows that $k(27) = h(13) = 13^{g(1)} = 13^{-1} = \frac{1}{13}$. As another example, we have

$$\begin{aligned}
k(196) &= h[F(196)] = h(-98) = -h(98) \\
&= -h(2^1 \cdot 7^2) \\
&= -2^{g(1)} \cdot 7^{g(2)} = -2^{-1} \cdot 7^1 \\
&= -\tfrac{7}{2}
\end{aligned}$$

Suppose now that we want to determine the unique positive integer which maps onto a given rational number under the mapping k; for instance, suppose we want to find $n \in I$ such that $k(n) = \frac{63}{64}$. Since k is a bijection on I to Q, we know that k^{-1} is a bijection on Q to I. Thus our task is to find $n = k^{-1}(\frac{63}{64})$. It follows from Prob. 10.3a that $k^{-1} = F^{-1} \circ h^{-1}$. Also, writing $i = i_Z$,

$$\begin{aligned}
\frac{63}{64} = \frac{3^2 \cdot 7^1}{2^6} &= 2^{-6} \cdot 3^2 \cdot 7^1 \\
&= 2^{i(-6)} \cdot 3^{i(2)} \cdot 7^{i(1)} \\
&= 2^{g[g^{-1}(-6)]} \cdot 3^{g[g^{-1}(2)]} \cdot 7^{g[g^{-1}(1)]} \\
&= 2^{g(11)} \cdot 3^{g(4)} \cdot 7^{g(2)}
\end{aligned}$$

and hence we have

$$k^{-1}(\tfrac{63}{64}) = F^{-1}[h^{-1}(\tfrac{63}{64})]$$
$$= F^{-1}[h^{-1}(2^{g(11)} \cdot 3^{g(4)} \cdot 7^{g(2)})]$$
$$= F^{-1}[h^{-1}\{h(2^{11} \cdot 3^4 \cdot 7^2)\}]$$
$$= F^{-1}(2^{11} \cdot 3^4 \cdot 7^2)$$
$$= 2(2^{11} \cdot 3^4 \cdot 7^2) + 1$$
$$= 2^{12} \cdot 3^4 \cdot 7^2 + 1$$
$$= 16{,}257{,}025$$

PROBLEMS

10.1 Prove Theorem 10.3.

10.2 Given f a bijection on H to K and g a bijection on K to H, prove:

(a) $\qquad\qquad\qquad\qquad\qquad g = f^{-1}$ $\qquad$ iff $g \circ f = i_H$

(b) $\qquad\qquad\qquad\qquad\qquad g = f^{-1}$ $\qquad$ iff $f = g^{-1}$

(c) $\qquad\qquad\qquad\qquad\qquad f = f^{-1}$ $\qquad$ iff $K = H$ and $f \circ f = i_H$

10.3 Let f be a bijection on A to B, g a bijection on B to C, and h a bijection on C to D. Define $k = g \circ f$ and $m = h \circ g \circ f$, recalling that composition of mappings is associative.

(a) Show that $f^{-1} \circ g^{-1} \circ k = i_A$, and conclude that $k^{-1} = f^{-1} \circ g^{-1}$.

(b) Show that m is a bijection on A to D and that $f^{-1} \circ g^{-1} \circ h^{-1} \circ m = i_A$. Conclude that $m^{-1} = f^{-1} \circ g^{-1} \circ h^{-1}$.

(c) Obtain the conclusion in (b) by writing $m = h \circ k$ and using the results in (a).

10.4 Verify that if f is a bijection on A to B, then f is a subset of $A \times B$ satisfying the following conditions:

(a) For each $a \in A$ there is a point $b \in B$ such that $\langle a,b \rangle \in f$.

(b) If $\langle a,b_1 \rangle \in f$ and $\langle a,b_2 \rangle \in f$, then $b_1 = b_2$.

(c) For each $b \in B$ there is a point $a \in A$ such that $\langle a,b \rangle \in f$.

(d) If $\langle a_1,b \rangle \in f$ and $\langle a_2,b \rangle \in f$, then $a_1 = a_2$.

Verify further that f^{-1} is the unique subset of $B \times A$ defined by

$$f^{-1} = \{\langle b,a \rangle \mid \langle a,b \rangle \in f\}$$

10.5 Verify that $g^{-1} \circ g = i_H$ and $g \circ g^{-1} = i_Z$ in Example 10.6.

10.6 Complete the details in arriving at the definition of F^{-1} in Example 10.7.

10.7 Prove that the mapping h defined in the proof of Theorem 10.8 is a bijection on Z to Q.

10.8 Using the mapping k defined in Theorem 10.8, find $k(40)$ and $k^{-1}(\tfrac{13}{18})$.

10.9 Prove that the composition of two increasing mappings is increasing and the composition of two decreasing mappings is increasing.

10.10 Let f be defined as follows:

$$f(x) = \begin{cases} x & \text{for } 0 \leqslant x < 1 \text{ and for } 2 \leqslant x < 3 \\ -x & \text{for } 1 \leqslant x < 2 \text{ and for } 3 \leqslant x < 4 \end{cases}$$

and in general, if n is a positive integer,

$$f(x) = (-1)^{n+1}x \qquad \text{for } n - 1 \leqslant x < n$$

Answer the following.   (It may help to sketch the graph.)

    (*a*) What is the domain of f?   Call it H.

    (*b*) What is the range of f?   Call it K.

    (*c*) Is f strictly monotone?

    (*d*) Is f one-to-one?

    (*e*) Does f^{-1} exist?   If not, why not?   If so, (i) state the domain and range of f^{-1} (ii) Try to formulate a definition of f^{-1} and verify that $f^{-1} \circ f$ is the identity mapping on H and $f \circ f^{-1}$ is the identity mapping on K.

10.11   Determine which of the mappings of Prob. 7.17 are strictly monotone.

10.12   Prove that if f is increasing, then f^{-1} is increasing, and if f is decreasing, then f^{-1} is decreasing.

10.13   For $p \in R_1$, define $M_p = R_1 - \{p\}$ and $g_p : M_p \to R_1$ by $g_p(x) = 1/(x - p)$ for each $x \in M_p$.   (Each mapping g_p is continuous on its domain M_p, by Corollary 9.11, and is a bijection on M_p to M_0, by Prob. 9.6.)

    (*a*) Denoting the inverse of g_p by the bijection h_p on M_0 to M_p, solve the equation $g_p(x) = y$ to show that the explicit formula for the mapping h_p is

$$h_p(y) = p + \frac{1}{y} \qquad \text{for each } y \in M_0$$

    (*b*) Show that the mapping g_p is decreasing on each of the open rays $\{x \mid x < p\}$ and $\{x \mid x > p\}$.

    (*c*) Show that g_p is unbounded on every deleted open interval (a,p,b) about p.

    (*d*) If (a,p,b) is any deleted open interval about p, show that

$$g_p\{(a,p)\} = \left\{ y \in R_1 \mid y < \frac{1}{a-p} \right\}$$

$$g_p\{(p,b)\} = \left\{ y \in R_1 \mid y > \frac{1}{b-p} \right\}$$

11

Countable Sets; Introduction to Sequences

Mappings defined on the set I, and the ranges of such mappings, are sufficiently important to warrant special terminology.

Definition 11.1

(a) A *sequence* is a mapping whose domain is the set I of positive integers.

(b) A set K is said to be *countable iff* $K = \varnothing$ or there is a sequence f such that $K = \{f\}$. A set which is not countable is called *uncountable*.

It follows from (b) above that a nonempty set K is countable iff there exists a mapping $f : I \xrightarrow{\text{onto}} K$. Several results are then immediate from our previous work.

Theorem 11.2

If A is countable and $f : A \xrightarrow{\text{onto}} B$, then B is countable.

Proof Since A is countable, there exists a mapping $g: I \xrightarrow{\text{onto}} A$ and
hence $f \circ g$ is a mapping of I onto B, by Theorem 7.22.   □

We sometimes state Theorem 11.2 in the form "onto mappings
preserve countability"; i.e., if the domain of an onto mapping is countable, its range is countable.

Corollary 11.3

Every subset of a countable set is countable.

Proof Every nonempty subset of a countable set A is a retract of A by
Theorem 7.27.   □

Our next theorem lists for easy reference several frequently oc-
curring countable sets. The proof of (a) is immediate from Theorem
10.8, and the rest follow by Corollary 11.3.

Theorem 11.4

Each of the following sets is countable:

(a) The set of all rational numbers
(b) The set of all integers
(c) The set of all nonnegative integers
(d) The set of all positive integers
(e) The empty set

A rather unusual application of Corollary 11.3 is illustrated in the
following example.

Example 11.5

Recalling that two integers are *relatively prime* iff they have no
common factor greater than 1, we see that the integers 3 and 5 of this
example could be any pair of relatively prime integers greater than 1.
We define K to be the set of all positive integers x such that x can be
written in the form $x = 3^h 5^k$, where h and k are nonnegative integers.
Since K is a subset of the countable set I, we have

(a) K is countable.

We leave as an exercise the proof that

(b) If $x = 3^h 5^k$ and $y = 3^r 5^s$ are points of K, then

$$x = y \qquad \text{iff } h = r \text{ and } k = s$$

Now let B be the set of all ordered pairs $\langle h,k \rangle$ of nonnegative integers. If for each $x = 3^h 5^k$ in K we define $f(x) = \langle h,k \rangle$, then it follows that

(c) f is a bijection on K to B.
(d) B is countable.

Defining H as the subset of B consisting of all ordered pairs $\langle h,k \rangle$ of positive integers, we have at once

(e) H is countable.

We summarize and extend the results of Example 11.5 in the next theorem. The proofs of (c) and (d) are left as exercises, following the same procedure as above using more than two relatively prime integers.

Theorem 11.6

The following sets are countable:

(a) The set of all ordered pairs $\langle h,k \rangle$ of nonnegative integers
(b) The set of all ordered pairs $\langle h,k \rangle$ of positive integers
(c) The set of all ordered triples $\langle h,k,t \rangle$ of nonnegative integers
(d) The set of all ordered n-tuples $\langle h_1,h_2, \ldots ,h_n \rangle$ of nonnegative integers

If K is a nonempty countable set, we know there exists a sequence f such that $\{f\} = K$. Since the domain of f is I, we may write the elements of K as $k_1 = f(1)$, $k_2 = f(2)$, $\ldots$, $k_n = f(n)$, $\ldots$. It then follows that

$$\{f\} = \{k_1,k_2, \ldots\} = \{k_n \mid n \in I\} = K \tag{1}$$

According to the definition given below, we may thus assert that K has been indexed by the set I. First however, let us generalize the above procedure by replacing the set K of points with a collection $\mathscr{K}$ of sets. If the nonempty collection $\mathscr{K}$ is countable, there is a sequence g such that $\{g\} = \mathscr{K}$. Writing $K_1 = g(1)$, $K_2 = g(2)$, $\ldots$, $K_n = g(n)$, $\ldots$, we have

$$\{g\} = \{K_1,K_2, \ldots\} = \{K_n \mid n \in I\} = \mathscr{K} \tag{2}$$

and we assert that the collection $\mathscr{K}$ of sets has been indexed by the set I. We remark that (1) may be considered as a special case of (2), where each set K_i is a singleton $\{k_i\}$.

Definition 11.7

Let H be a nonempty set and $\mathcal{K}$ a collection of sets. If f is a mapping of H onto $\mathcal{K}$, and if for each point $\alpha \in H$ we denote the image of α under f in $\mathcal{K}$ by $f(\alpha) = K_\alpha$, then we say that the collection $\mathcal{K} = \{K_\alpha \mid \alpha \in H\}$ has been *indexed by the set H under the mapping f*. The mapping f is called the *index mapping*, and the set H is called the *index set*.

We often bypass the details in Definition 11.7 with a statement such as "let H be an index set, and for every $p \in H$, let A_p be a set." It is evident from this definition that nonempty countable sets are precisely those sets which can be indexed by the set I, with a sequence serving as the index mapping. The reader is cautioned against the indiscriminate use of I as an index set in general situations, however, since there are many uncountable sets. In particular, the index set H of our next definition and theorem may be uncountable.

Definition 11.8

Let H be an index set, and for each $p \in H$ let A_p be a set. Then:

(a) The *union of the collection of sets* $\{A_p \mid p \in H\}$ (written $\bigcup_{p \in H} A_p$) is the set consisting of all points x such that $x \in A_p$ for at least one $p \in H$.

(b) The *intersection of the collection of sets* $\{A_p \mid p \in H\}$ (written $\bigcap_{p \in H} A_p$) is the set consisting of all points x such that $x \in A_p$ for every $p \in H$.

The proof of our next theorem is an easy consequence of Definition 11.8 and is left as an exercise. Note that (c) and (d) generalize De Morgan's laws to arbitrary collections of sets.

Theorem 11.9

Let H be an index set, and for each $p \in H$ let A_p be a set. Then:

(a) If $q \in H$, then $A_q \subset \bigcup_{p \in H} A_p$.

(b) If $q \in H$, then $\bigcap_{p \in H} A_p \subset A_q$.

(c) $C(\bigcup_{p \in H} A_p) = \bigcap_{p \in H} C(A_p).$ $\left.\vphantom{\begin{array}{c} a \\ b \end{array}}\right\}$ De Morgan's laws

(d) $C(\bigcap_{p \in H} A_p) = \bigcup_{p \in H} C(A_p).$

Theorem   11.10

The union of any countable collection of countable sets is countable.

Proof   For each positive integer h let A_h be a countable set, and let J
   be the union of the sets A_h. We want to show that J is countable.
   If all the sets A_h are empty, then J is empty and hence countable.
   We may thus suppose that at least one of the sets is nonempty,
   and we lose no generality in assuming $A_1 \neq \varnothing$. Since each countable set A_h can be indexed by I, we use a double subscript notation, denoting by $a_{h,k}$ the kth term of the set A_h, where $k \in I$. It
   then follows that $a_{1,1} \in A_1$. Let K be the set of all ordered pairs
   $\langle h,k \rangle$ of positive integers, and define the mapping f as follows:

$$\text{For every } x = \langle h,k \rangle \text{ in } K \quad f(x) = \begin{cases} a_{h,k} & \text{if } a_{h,k} \in A_h \\ a_{1,1} & \text{if } a_{h,k} \notin A_h \end{cases}$$

It is clear that f is a mapping of K onto J. But K is countable
by Theorem 11.6b, and hence J is countable by Theorem
11.2.   □

To satisfy the curious reader, we remark that any open interval
(and hence any superset of an open interval) is uncountable, but we
defer the proof to a later chapter.

We have defined a sequence as a mapping with domain I. Thus
a sequence $f:I \to R_1$ is a subset of the Cartesian product $I \times R_1$, and
may be written as

$$f = \langle \langle 1,f(1) \rangle, \langle 2,f(2) \rangle, \ \ldots \ , \langle n,f(n) \rangle, \ \ldots \ \rangle,$$

where the outer angle brackets indicate that the ordered pairs $\langle n,f(n) \rangle$
are themselves ordered according to the natural ordering of their first
components. With this convention, it is evident that a sequence is
completely determined by its values at successive positive integers n,
and hence may be written in the abbreviated form

$$f = \langle f(1),f(2), \ \ldots \ ,f(n), \ \ldots \ \rangle$$

If we now define $f(n) = x_n$ for each $n \in I$, it follows that

$$f = \langle x_1,x_2, \ \ldots \ ,x_n, \ \ldots \ \rangle$$

which we often simplify further by writing merely $f = \langle x_n \rangle$. For each
$n \in I$, the real number $x_n = f(n)$ is called the nth *term* of the
sequence, and it should be clear that the set of all distinct terms of f
constitutes the range of f; that is, if $f = \langle x_n \rangle$, then $\{f\} = \{x_n\}$.

In writing a sequence as $f = \langle x_n \rangle$, we are conforming to a notational practice which has developed over many years. It amounts to
equating the sequence (which is a mapping) with its terms (which are

values of the mapping). No real harm is done so long as we fully understand the notation. Another long-established convention is the practice of referring to the terms of a sequence as a *sequence of points*, and we shall also use this terminology. Thus we may write "let $\langle p_n \rangle$ be a sequence of points in a set H," which really means "let $f : I \to H$ be a sequence defined by $f(n) = p_n$ for each $n \in I$" or "let $f = \langle p_n \rangle$, where $\{p_n\} \subset H$." Similarly, we may speak of a "sequence $\langle K_n \rangle$ of sets" in a collection $\mathscr{K}$, rather than a sequence $g : I \to \mathscr{K}$ defined by $f(n) = K_n$ for each $n \in I$. In general, any sequence f such that $\{f\} \subset R_1$ is called a real-valued sequence or, more briefly, a *real sequence*, and may also be referred to as a *sequence of real numbers*. A sequence which is one-to-one is often called a *sequence of distinct points*.

Since sequences are merely special mappings, most of the general results on mappings in previous chapters apply automatically to sequences. For example, a sequence $\langle x_n \rangle$ is bounded iff there exists a real number $M > 0$ such that $|x_n| \leqslant M$ for every $n \in I$. A sequence $\langle x_n \rangle$ is increasing iff $x_n < x_{n+1}$ for every $n \in I$, with corresponding statements for the other monotone sequences. The sum of the sequences $f = \langle x_n \rangle$ and $g = \langle y_n \rangle$ is the sequence $f + g = \langle x_n + y_n \rangle$. Similarly, $f - g = \langle x_n - y_n \rangle$, $fg = \langle x_n y_n \rangle$, and $f/g = \langle x_n/y_n \rangle$, provided $y_n \neq 0$ for $n \in I$. Note also that the composition $f \circ g$ is defined if $\{g\} \subset I$. We shall now see how the composition of f with a special sequence β produces a subsequence of f.

Let $f = \langle x_n \rangle$, and suppose that $\beta : I \to I$ is an increasing sequence; in other words, $\langle \beta(n) \rangle$ is an increasing sequence of positive integers. Then the composition $f \circ \beta$ defined on I is a sequence which we denote by $f \circ \beta = g = \langle y_n \rangle$. Note that for each $n \in I$ we thus have

$$y_n = g(n) = (f \circ \beta)(n) = f[\beta(n)] = x_{\beta(n)}$$

This means that the nth term of the sequence $\langle y_n \rangle$ is the $\beta(n)$th term of the sequence $\langle x_n \rangle$. Hence the increasing sequence β merely selects certain terms from the sequence $\langle x_n \rangle$ (and in the same order in which they occur in $\langle x_n \rangle$) to constitute the sequence $\langle y_n \rangle$. We thus say that $\langle y_n \rangle$ is a subsequence of $\langle x_n \rangle$.

Definition 11.11

A sequence $g = \langle y_n \rangle$ is said to be a *subsequence* of a sequence $f = \langle x_n \rangle$ iff there exists an increasing sequence $\beta : I \to I$ such that $g = f \circ \beta$. If such is the case, then $y_n = x_{\beta(n)}$ for every $n \in I$.

Since the identity sequence i is increasing and satisfies $f \circ i = f$ for every sequence f, it follows that any sequence is a subsequence of itself. For a less trivial example, let $f = \langle x_n \rangle$, and suppose that $\beta : I \to I$ is defined by $\beta(n) = 2n - 1$ for each $n \in I$. The reader may verify that β is increasing; thus letting $f \circ \beta = g = \langle y_n \rangle$, we have

$$\langle y_1, y_2, y_3, \ldots, y_n, \ldots \rangle = \langle x_1, x_3, x_5, \ldots, x_{2n-1}, \ldots \rangle$$

Note that since $\langle \beta(n) \rangle$ is the sequence of odd positive integers with their natural ordering, the subsequence $\langle y_n \rangle$ consists of the odd-numbered terms, i.e., the terms with odd subscripts, of the sequence $\langle x_n \rangle$.

Theorem 11.12

If $g = \langle y_n \rangle$ is a subsequence of the sequence $f = \langle x_n \rangle$, and if $h = \langle z_n \rangle$ is a subsequence of $\langle y_n \rangle$, then $\langle z_n \rangle$ is a subsequence of $\langle x_n \rangle$.

Proof Since g is a subsequence of f, there exists an increasing sequence $\beta : I \to I$ such that $g = f \circ \beta$, and since h is a subsequence of g, there exists an increasing sequence $\psi : I \to I$ such that $h = g \circ \psi$. We then have

$$h = g \circ \psi = (f \circ \beta) \circ \psi = f \circ (\beta \circ \psi),$$

using the associativity of compositions. The mapping $\beta \circ \psi$ is clearly a sequence, and it is increasing, by Prob. 10.9. Therefore $h = \langle z_n \rangle$ is a subsequence of $f = \langle x_n \rangle$. Note that $z_n = x_{\beta[\psi(n)]}$ for each $n \in I$. $\quad\square$

Theorem 11.13

If the range $\{f\}$ of a sequence f is a finite set, then f has a constant subsequence.

Proof Suppose that $K = \{f\}$ is a finite set. By Theorem 4.11, there is a positive integer t such that $K = \{x_1, x_2, \ldots, x_t\}$. Since f is a mapping of I onto K, for each integer i such that $1 \leq i \leq t$ we define the set

$$H_i = \{n \in I \mid f(n) = x_i\}$$

Clearly, $I = H_1 \cup H_2 \cup \ldots \cup H_t$, so that by Corollary 4.10, at least one of the sets H_i must be infinite. Let t_0 be the smallest positive integer such that H_{t_0} is an infinite subset of I. We may order the elements of H_{t_0} by the well-ordering property, so that $H_{t_0} = \{n_1, n_2, \ldots\}$, where $n_1 < n_2 < \ldots$. Defining $\beta : I \to I$ by $\beta(i) = n_i$ for each $i \in I$, it is clear that β is increasing, and hence $f \circ \beta$ is a subsequence of f. Furthermore, for each $n \in I$, $(f \circ \beta)(n) = f[\beta(n)] = f(H_{t_0}) = x_{t_0}$. $\quad\square$

Theorem 11.14

Every sequence of real numbers has a monotone subsequence.

Proof Let $f = \langle x_n \rangle$ be any sequence of real numbers. If the range $\{f\}$ of f is a finite set, then f has a constant (and hence monotone) subsequence, by Theorem 11.13. Thus we may suppose that $\{f\}$ is infinite. If $\{f\}$ is bounded, then there exists a nonempty subset K of $\{f\}$ such that sup $K \notin K$ or inf $K \notin K$. If $\{f\}$ is unbounded, then $\{f\}$ cannot have both a supremum and an infimum, and so we may choose $K = \{f\}$.

Suppose that no member of K is a supremum of K. Let i_1 be the smallest positive integer such that $f(i_1) \in K$. Since none of the real numbers $f(n)$, where $1 \leqslant n \leqslant i_1$, can be a supremum of K, define i_2 as the smallest integer greater than i_1 such that $f(i_2) \in K$ and $f(i_1) < f(i_2)$. If $f(i_n)$ has been defined for $1 \leqslant n \leqslant k$ and as satisfying $f(i_1) < f(i_2) < \cdots < f(i_k)$, define i_{k+1} as the smallest integer greater than i_k such that $f(i_{k+1}) \in K$ and $f(i_k) < f(i_{k+1})$. We have thus defined inductively an increasing sequence $\beta : I \to I$, where $\beta(n) = i_n$ for each $n \in I$, and it is clear that $f \circ \beta$ is an increasing subsequence of f.

In the case where no member of K is an infinum of K, a similar inductive procedure yields a decreasing subsequence of f. $\square$

The above proof suggests another convenient notation for subsequences that is often used, although it tends to obscure the mappings involved. We have seen that if $g = f \circ \beta = \langle y_n \rangle$ is a subsequence of the sequence $f = \langle x_n \rangle$, then $y_n = x_{\beta(n)}$ for each $n \in I$. If, as in the proof above, we denote the increasing mapping $\beta : I \to I$ by $\beta(n) = i_n$ for each $n \in I$, it follows that $y_n = x_{i_n}$ for each $n \in I$. We may thus speak of the subsequence $\langle x_{i_n} \rangle$ of the sequence $\langle x_n \rangle$.

PROBLEMS

11.1 If f is a mapping of A onto B and B is uncountable, is A uncountable? Why?

11.2 Explain why the following are countable sets:
(a) The set of all pairs of rational numbers
(b) The collection of all line segments of R_1 which have rational length and rational midpoints
(c) The collection of all circles in the plane which have rational centers and rational radii

11.3 Prove assertion (b) of Example 11.5.

11.4 Give a simple example to show that assertion (b) of Example 11.5 is not true if the integers (we used 3 and 5) are not relatively prime. Explain how this would affect our justification of assertions (c), (d), and (e) of Example 11.5.

11.5 Prove (c) and (d) of Theorem 11.6.

11.6 Prove Theorem 11.9.

11.7 Let H be an index set, and for each $\alpha \in H$ let A_α be a set. Prove that for any set K,

(a)
$$\left(\bigcap_{\alpha \in H} A_\alpha \right) \cap K = \bigcap_{\alpha \in H} (A_\alpha \cap K)$$

(b)
$$\left(\bigcup_{\alpha \in H} A_\alpha \right) \cup K = \bigcup_{\alpha \in H} (A_\alpha \cup K)$$

(c)
$$\left(\bigcup_{\alpha \in H} A_\alpha \right) \cap K = \bigcup_{\alpha \in H} (A_\alpha \cap K)$$

(d)
$$\left(\bigcap_{\alpha \in H} A_\alpha \right) \cup K = \bigcap_{\alpha \in H} (A_\alpha \cup K)$$

In the remaining problems, H is a nonempty subset of R_1, p is a point of R_1, $H_1 = \{x \in H \mid x < p\}$, and $H_2 = \{x \in H \mid x > p\}$. We shall also refer to the following sequences of sets. For each $n \in I$,

$$A_n = H \cap (n, n+1]$$
$$B_n = H \cap [-n-1, -n)$$
$$C_n = H \cap \left[p - \frac{1}{n}, p - \frac{1}{n+1} \right)$$
$$D_n = H \cap \left(p + \frac{1}{n+1}, p + \frac{1}{n} \right]$$

11.8 Prove that H has no upper bound iff given any positive integer N, there is an integer $n > N + 1$ such that $A_n \neq \varnothing$.

11.9 Prove that H has no lower bound iff given any positive integer N, there is an integer $n > N + 1$ such that $B_n \neq \varnothing$.

11.10 Suppose that H has no upper bound. Recalling the well-ordering property of I, and using the result of Prob. 11.8, let i_1 be the smallest positive integer such that $A_{i_1} \neq \varnothing$. Now define a subsequence $\langle A_{i_n} \rangle$ of the sequence $\langle A_n \rangle$ by mathematical induction. If A_{i_k} has been defined, let i_{k+1} be the smallest integer greater than $1 + i_k$ such that $A_{i_{k+1}} \neq \varnothing$. Finally, for each $n \in I$, define $K_n = A_{i_n}$. Prove that the sequence $\langle K_n \rangle$ has the following properties:

    (a) K_n is a nonempty subset of H for every $n \in I$.
    (b) If $x \in K_n$ and $y \in K_{n+1}$, then $x + 1 < y$.

11.11 Suppose that H has no lower bound. Prove that there is a sequence $\langle K_n \rangle$ of nonempty subsets of H such that if $x \in K_n$ and $y \in K_{n+1}$, then $y < x - 1$.

11.12 Prove that p is a cluster point of H_1 iff there is a subsequence $\langle C_{i_n} \rangle$ of $\langle C_n \rangle$ such that $C_{i_n} \neq \varnothing$ for every $n \in I$.

11.13 Prove that p is a cluster point of H_2 iff there is a subsequence $\langle D_{i_n} \rangle$ of $\langle D_n \rangle$ such that $D_{i_n} \neq \varnothing$ for every $n \in I$.

12
Sequences and Cluster Points; Convergence; Cauchy Sequences

We now introduce an additional axiom which is a useful tool in our work with sequences. It is a special case of a more general axiom known in mathematics as the *axiom of choice.*

Axiom of choice for sequences

If $\langle A_n \rangle$ is a sequence of nonempty sets, then there exists a sequence of points $\langle p_n \rangle$ such that $p_n \in A_n$ for each $n \in I$.

Definition 12.1

We shall call a sequence f a *uniformly isolated sequence* iff f is one-to-one and $\{f\}$ is a uniformly isolated set.

Theorem 12.2

A subset H of R_1 is unbounded iff it contains a monotone sequence which is uniformly isolated.

Proof We suppose first that H is unbounded and consider the case in which H has no upper bound, leaving the case in which H has no lower bound as an exercise. By Prob. 11.10, there is a sequence $\langle K_n \rangle$ of nonempty subsets of H such that if $x \in K_n$ and $y \in K_{n+1}$, then $x + 1 < y$. By the axiom of choice for sequences, there is a sequence $\langle p_n \rangle$ such that $p_n \in K_n$ for each $n \in I$. Clearly

$$p_n + 1 < p_{n+1} \qquad \text{for each } n \in I$$

so that $\langle p_n \rangle$ is an increasing sequence of points of H which is uniformly isolated.

For the converse, we suppose that $\langle x_n \rangle$ is an increasing uniformly isolated sequence of points of H and leave the case where $\langle x_n \rangle$ is decreasing as an exercise. There exists a real number $r > 0$ such that $x_{n+1} - x_n > r$ for each $n \in I$. Hence we have

$$x_2 > x_1 + r$$
$$x_3 > x_2 + r > x_1 + 2r$$

and in general, for each positive integer n,

$$x_{n+1} > x_n + r > x_1 + nr$$

For any $M > 0$ we may choose $n \in I$ by the Archimedean property so that $x_1 + nr > M$. Since $x_{n+1} \in H$, it follows that H has no upper bound. $\square$

Theorem 12.3

A point p is a cluster point of a set H iff every deleted open interval about p contains a strictly monotone sequence of points of H.

Proof The "if" part of the theorem is a simple exercise. To prove the "only if" part, suppose that p is a cluster point of H, and let (a,p,b) be any deleted open interval about p. We shall use the terminology introduced in Probs. 11.8 to 11.13. By Prob. 6.10, p is a cluster point of at least one of the sets H_1, H_2, and we can choose our notation so that $p \in H_1'$. Using the subsequence $\langle C_{i_n} \rangle$ of nonempty subsets of H determined by Prob. 11.12 we see that the axiom of choice for sequences guarantees the existence of a sequence $\langle p_n \rangle$ such that $p_n \in C_{i_n}$ for each $n \in I$. Choosing N by the Archimedean property so that $1/N < p - a$, we can be certain that $p_n \in (a,p) \subset (a,p,b)$ for every $n \geq N$. If we let $f = \langle p_n \rangle$ and define $\beta: I \to I$ by $\beta(n) = N + n$ for each $n \in I$, then $f \circ \beta$ fulfills the requirements of the theorem. $\square$

Corollary 12.4

A nonempty bounded subset of R_1 is infinite iff it contains a strictly monotone sequence of points.

This corollary is immediate by virtue of the Bolzano-Weierstrass theorem (Theorem 6.10). Before listing some additional corollaries to Theorem 12.3, we introduce a definition which will simplify their statements.

Definition 12.5

A sequence $\langle p_n \rangle$ is said to *converge* to the point p iff given any real number $\epsilon > 0$, there exists a positive integer N such that $|p_n - p| < \epsilon$ for every integer $n \geqslant N$. The point p is called a *limit* of the sequence $\langle p_n \rangle$, and we write $\lim p_n = p$. If H is a nonempty subset of R_1, then a sequence of points of H is said to be *convergent in H* iff there exists a point $p \in H$ to which it converges. A sequence which is convergent in R_1 will simply be called *convergent*.

Corollary 12.6

A point p is a cluster point of a set H iff there exists a strictly monotone sequence of points of H which converges to p.

Corollary 12.7

If $p = \sup H$ and $p \notin H$, then there is an increasing sequence of points of H which converges to p.

Corollary 12.8

If $p = \inf H$ and $p \notin H$, then there is a decreasing sequence of points of H which converges to p.

We now return to a discussion of Definition 12.5 and some of its consequences. Inasmuch as a sequence is a mapping, it seems reasonable to desire a characterization of a convergent sequence based purely on the concept of a mapping. Suppose then that $f: I \rightarrow R_1$ is a sequence. Defining $f(n) = x_n$ for each $n \in I$, we may write $f = \langle x_n \mid n \in I \rangle$ and $\{f\} = \{x_n \mid n \in I\}$. We introduce a special class of subsequences of f which we shall call *truncated sequences*. For each positive integer N, the truncated sequence f_N is defined by

$f_N = \langle x_n \mid n \geqslant N \rangle$. Thus

$$f_1 = \langle x_1, x_2, x_3, \ldots \rangle$$
$$f_2 = \langle x_2, x_3, x_4, \ldots \rangle$$
$$\cdots \cdots \cdots \cdots \cdots$$
$$f_N = \langle x_N, x_{N+1}, x_{N+2}, \ldots \rangle$$
$$\cdots \cdots \cdots \cdots \cdots$$

With this convenient notation, the following characterization of a convergent sequence is merely a rewording of Definition 12.5.

Theorem 12.9

A sequence f converges to a point p iff given any open interval (a,b) containing p there exists a positive integer N such that $\{f_N\} \subset (a,b)$.

Corollary 12.10

A convergent sequence has a unique limit; i.e., if f converges to p, and if $q \neq p$, then f does not converge to q.

Proof By the Hausdorff property (see Prob. 5.9), there exist disjoint open intervals (a,b) and (c,d) such that $p \in (a,b)$ and $q \in (c,d)$. Since f converges to p, there is a positive integer N such that $\{f_N\} \subset (a,b)$. Thus for any positive integer $N_1 \geqslant N$, it follows that $\{f_{N_1}\} \cap (c,d) = \varnothing$, and hence f does not converge to q.   □

Corollary 12.11

Every subsequence of a convergent sequence f is convergent and converges to the limit of f.

Proof The reader may easily verify that if $g = f \circ \beta$ is a subsequence of f, then given any positive integer N, there exists a positive integer N_1 such that $\{g_{N_1}\} \subset \{f_N\}$.

Corollary 12.12

Every convergent sequence is bounded.

Proof Let f converge to p, and let (a,b) be an open interval containing p. Then there exists a positive integer N such that $\{f_N\} \subset (a,b)$. Defining

$$M = \sup \{|a|, |b|, |f(1)|, |f(2)|, \ldots, |f(N-1)|\}$$

it follows that $\{f\} \subset (-M, M)$, and hence f is bounded.   □

Corollary 12.13

If a sequence $f = \langle x_n \rangle$ converges to a point p, and if (a,b) is any open interval containing p, then there exists a positive integer N such that if $h \geqslant N$ and $k \geqslant N$, then $|x_h - x_k| < b - a$.

Corollary 12.14

If a sequence f converges to a point p and N is a positive integer such that $\{f_N\} \subset (c,d)$, then $p \in [c,d]$.

Proof If $d < p$, then $p \in (d, p + 1)$ and $(d, p + 1) \cap \{f_N\} = \varnothing$, with a similar result holding if $p < c$. In either case, f cannot converge to p. □

We remark that Corollary 12.11 provides a useful method for showing that a sequence f is not convergent, since it suffices to display two subsequences of f which converge to different limits. For example, the sequence f defined by $f(n) = (-1)^n$ for each $n \in I$ is not convergent since the subsequence of even terms converges to 1 and the subsequence of odd terms converges to -1. Thus f is a bounded sequence which does not converge, and so the converse of Corollary 12.12 is not true. It follows that boundedness is a necessary but not a sufficient condition for convergence of a sequence. However, boundedness *plus* monotonicity is sufficient, as shown by the following theorem.

Theorem 12.15

Every bounded monotone sequence f is convergent; in fact, if f is nondecreasing, then f converges to sup $\{f\}$, and if f is nonincreasing, then f converges to inf $\{f\}$.

Proof We suppose that the bounded sequence f is nondecreasing, and leave the similar case where f is nonincreasing as an exercise. Let $p = \sup \{f\}$ and let (a,b) be any open interval containing p. Then a is not an upper bound for $\{f\}$, and so there exists a positive integer N such that $f(N) \in (a,p]$. Since f is nondecreasing, it follows that $a < f(N) \leqslant f(n) \leqslant p$ for all integers $n \geqslant N$ and hence $\{f_N\} \subset (a,b)$. □

Combining Theorem 12.15 and the contrapositive of Corollary 12.12 yields the following convergence criterion for monotone sequences.

Theorem 12.16

A monotone sequence of real numbers is convergent iff it is bounded.

Corollary 12.17   The Bolzano-Weierstrass theorem for sequences

Every bounded sequence of real numbers has a convergent sub-sequence.

Proof   Every bounded sequence of real numbers has a monotone subsequence by Theorem 11.14; this subsequence is also bounded and hence convergent by Theorem 12.16. $\square$

We shall now derive a characterization of a convergent sequence of real numbers which does not depend on the value of its limit. Note this one disadvantage (that we must somehow determine the value of the limit before proving convergence) in the use of Definition 12.5 or the equivalent characterization in Theorem 12.9.

Definition 12.18

Let H be a nonempty subset of R_1. A sequence $f = \langle x_n \rangle$ of points of H is said to be a *Cauchy sequence in H* iff given any $\epsilon > 0$, there exists a positive integer N such that if $h \geqslant N$ and $k \geqslant N$, then $|x_h - x_k| < \epsilon$. A Cauchy sequence in R_1 will simply be called a *Cauchy sequence*.

Theorem 12.19

(a) A sequence f is a Cauchy sequence iff given any $\epsilon > 0$, there exist a positive integer N and real numbers a, b with $0 < b - a \leqslant \epsilon$ such that $\{f_N\} \subset (a,b)$.
(b) Every subsequence of a Cauchy sequence is a Cauchy sequence.
(c) Every Cauchy sequence is bounded.
(d) Every Cauchy sequence has a convergent subsequence.
(e) Every convergent sequence is a Cauchy sequence.

Proof   We offer the following suggestions, leaving the details as an exercise. For the "only if" part of (a), given a Cauchy sequence f and $\epsilon > 0$, choose $a = f(N) - \epsilon/2$ and $b = f(N) + \epsilon/2$, where the positive integer N is determined from Definition 12.18. For the "if" part of (a), given a sequence f and $\epsilon > 0$, use the hypothesis to select N, a, b with $0 < b - a \leqslant \epsilon/2$ such that $\{f_N\} \subset (a,b)$, and then show that for $h \geqslant N$ and $k \geqslant N$,

$$|x_h - x_k| = |x_h - x_N + x_N - x_k| \leqslant |x_h - x_N| + |x_N - x_k| < \frac{\epsilon}{2} + \frac{\epsilon}{2} = \epsilon$$

For (b), (c), and (e), see Corollaries 12.11, 12.12, and 12.13, respectively; (d) follows from (c) and Corollary 12.17.

Theorem 12.20

Every Cauchy sequence in R_1 is convergent in R_1.

Proof Let f be a Cauchy sequence in R_1. By Theorem 12.19d, f has a convergent subsequence g; so suppose g converges to p. We shall show that f also converges to p. Let (a,b) be any open interval containing p, and define $\epsilon = \inf \{(p-a)/2, (b-p)/2\}$. Since f is a Cauchy sequence, there exists a positive integer N and real numbers c, d with $0 < d - c \leqslant \epsilon$ such that $\{f_N\} \subset (c,d)$. Since g is a subsequence of f, there exists a positive integer N_1 such that $\{g_{N_1}\} \subset \{f_N\} \subset (c,d)$. But g converges to p, and hence $p \in [c,d]$, from which it easily follows that $[c,d] \subset (a,b)$. Therefore $\{f_N\} \subset (a,b)$. $\square$

Combining Theorems 12.19e and 12.20, we have the desired characterization.

Theorem 12.21   Cauchy convergence criterion

A sequence in R_1 is convergent iff it is a Cauchy sequence.

The importance of Cauchy sequences and the Cauchy convergence criterion is perhaps not too apparent at our present level of development. The student might argue that since every convergent sequence in R_1 is Cauchy and every Cauchy sequence in R_1 is convergent, why bother to distinguish between these two concepts? We have already stated one good reason—that convergence of a sequence in R_1 can be established without considering the value of the limit—and it has implications which lie deep in analysis. Some hint of these further implications is given below and in a later chapter.

Let us consider the open interval $H = (0,2)$ and the sequence $f = \langle 1/n \rangle$ of points of H. It is easy to see that f is a Cauchy sequence in H which is not convergent in H. (Of course, f is both Cauchy and convergent, with limit 0, in R_1). According to the following definition, we say that H is not complete.

Definition 12.22

A subset H of R_1 is said to be *complete* iff every Cauchy sequence in H is convergent in H.

The Cauchy convergence criterion asserts that R_1 is complete. It is no accident that we referred to the least upper bound axiom as the completeness axiom, since the property of completeness as stated in Definition 12.22 and the completeness axiom are equivalent.

PROBLEMS

12.1   Prove that every subsequence of a uniformly isolated sequence is uniformly isolated.

12.2   State precisely what is meant by the assertion that a sequence $f = \langle x_n \rangle$ of real numbers is not uniformly isolated.

12.3   Prove that every subsequence of a Cauchy sequence is a Cauchy sequence.

12.4   State precisely what is meant by the assertion that a sequence $f = \langle x_n \rangle$ of real numbers is not a Cauchy sequence.

12.5   Give an example of a sequence which is neither Cauchy nor uniformly isolated.

12.6   Complete the proof of Theorem 12.2 by considering the cases omitted in the text.

12.7   Prove the "if" part of Theorem 12.3.

12.8   Justify the statements of Corollaries 12.6, 12.7, 12.8, and 12.13.

12.9   Show that if the limit p of a convergent sequence f is not a cluster point of $\{f\}$, then there exists a positive integer N such that $\{f_N\} = \{p\}$.

12.10   Prove the case where f is nonincreasing in Theorem 12.15.

12.11   Complete the details of the proof of Theorem 12.19.

12.13   Prove that a set H is infinite iff H contains a sequence of distinct points; that is, H is infinite iff there exists a one-to-one mapping $f : I \rightarrow H$.

12.14   Prove that if a sequence $f = \langle x_n \rangle$ contains a uniformly isolated subsequence, then f is not convergent.

12.15   For a sequence $f = \langle x_n \rangle$, the sequence $|f|$ is defined in Prob. 9.3 by $|f|(n) = |x_n|$ for each $n \in I$. Prove that if $\lim x_n = p$, then $\lim |x_n| = |p|$. (*Hint:* Use Theorem 3.12i.) Is the converse true? If so, prove it; if not, give a counterexample.

12.16   Let $f = \langle x_n \rangle$, $g = \langle y_n \rangle$, and $h = \langle z_n \rangle$ be sequences such that $x_n \leqslant y_n \leqslant z_n$ for each $n \in I$. Suppose that f and h are convergent, with $\lim x_n = p$ and $\lim z_n = p$. Prove that g is convergent and $\lim y_n = p$.

12.17   Prove that $\lim |x_n| = 0$ iff $\lim (-|x_n|) = 0$.

12.18   Prove that if $\lim |x_n| = 0$, then $\lim x_n = 0$. *Hint:* Use Probs. 12.16 and 12.17.

12.19   Prove that if $|r| < 1$, then $\lim r^n = 0$. *Hint:* Use Theorem 12.15 on $|r|^n$ and then apply Prob. 12.18.

12.20   Let $f = \langle p_n \rangle$ be a sequence of distinct points in R_1 and $q \in R_1$ be a cluster point of $\{f\}$. Prove that f has a subsequence converging to q.

13

Infinite Series; Sets of Measure Zero; Uncountable Sets; Some Characterizations of Finite and Infinite Sets

Although an exhaustive study of infinite series is not our intention in this text, this important subject is a necessary prelude to our discussion of sets of measure zero. We develop the theory only to the extent required by our subsequent work and refer readers interested in a more thorough treatment to any of the several texts on advanced calculus or real analysis. A particularly nice treatment appears in "Methods of Real Analysis" by R. R. Goldberg (Blaisdell, 1964), and we shall adopt his approach in defining an infinite series as an ordered pair of related sequences.

Associated with any sequence $f = \langle x_n \rangle$ of real numbers is another uniquely determined sequence $g = \langle s_n \rangle$ of real numbers defined by

$$s_n = x_1 + x_2 + \cdots + x_n \qquad \text{for each } n \in I$$

This association is reciprocal in that g also uniquely determines f, since

$$x_1 = s_1 \qquad \text{and} \qquad x_n = s_n - s_{n-1} \qquad \text{for } n > 1, n \in I$$

It follows that the ordered pair $\langle f,g \rangle$ whose first component is the sequence $f = \langle x_n \rangle$ and whose second component is the sequence $g = \langle s_n \rangle$ is completely determined by specifying either component. According to the definition which we now give, this ordered pair of sequences of real numbers is called an infinite series of real numbers, and may be denoted by $\sum_{n \in I} x_n$, or more briefly, by Σx_n. Sometimes it is convenient to indicate the series by $x_1 + x_2 + x_3 + \cdots$.

Definition 13.1

An *infinite series* $\sum_{n \in I} x_n$ is an ordered pair $\langle f,g \rangle$ of sequences where $f = \langle x_n \rangle$, and $g = \langle s_n \rangle$ is defined by

$$s_n = x_1 + x_2 + \cdots + x_n \qquad \text{for each } n \in I$$

For each $n \in I$, x_n is called the *n*th *term* of the series, and s_n is called the *n*th *partial sum* of the series, so that $g = \langle s_n \rangle$ is the *sequence of partial sums* of the series.

Definition 13.2

An infinite series Σx_n is said to be *convergent* iff its sequence $\langle s_n \rangle$ of partial sums is convergent. If $\langle s_n \rangle$ is not convergent, then the series Σx_n is said to be *divergent*. If Σx_n is a convergent series with $\lim s_n = S$, then we say that the series *converges to S* and call S the *sum* of the series.

We may thus talk about the sum of a convergent series, even though the series consists of an infinite number of summands. When a series Σx_n converges to a real number S, we often write $\Sigma x_n = S$, in which case the symbols Σx_n represent both the infinite series and its sum.

Our first easy but useful theorem gives a *necessary condition for convergence* of an infinite series.

Theorem 13.3

If Σx_n is convergent, then $\lim x_n = 0$.

Proof The convergence of Σx_n implies convergence of the sequence $\langle s_n \rangle$ of partial sums. Then $\langle s_n \rangle$ is a Cauchy sequence, from which it follows that $\lim (s_n - s_{n-1}) = 0$. But $x_n = s_n - s_{n-1}$ for each n, and hence $\lim x_n = 0$. $\square$

The contrapositive of Theorem 13.3 provides a *sufficient condition for divergence* of an infinite series and is sometimes called the n*th term test for divergence.* It asserts that if the terms x_n of the series Σx_n do not approach zero as n increases, then the series is divergent. We emphasize, however, that $\lim x_n = 0$ is not a sufficient condition for convergence of Σx_n, and illustrate with an example.

Example 13.4

Consider the *harmonic series* $\Sigma(1/n)$ which obviously satisfies the condition $\lim (1/n) = 0$. We shall show that this series is divergent by proving that its sequence $\langle s_n \rangle$ of partial sums is not convergent. Note that

$$s_2 - s_1 = \tfrac{1}{2}$$
$$s_4 - s_2 = \tfrac{1}{3} + \tfrac{1}{4} > \tfrac{1}{4} + \tfrac{1}{4} = \tfrac{1}{2}$$
$$s_8 - s_4 = \tfrac{1}{5} + \tfrac{1}{6} + \tfrac{1}{7} + \tfrac{1}{8} > \tfrac{1}{8} + \tfrac{1}{8} + \tfrac{1}{8} + \tfrac{1}{8} = \tfrac{1}{2}$$

Similarly,

$$s_{16} - s_8 > 8 \left(\frac{1}{16} \right) = \frac{1}{2}$$

and in general for any $k \in I$

$$s_2{}^k - s_2{}^{k-1} > 2^{k-1} \frac{1}{2^k} = \frac{1}{2}$$

Thus the subsequence $\langle s_{2^n} \rangle$ of the sequence $\langle s_n \rangle$ is uniformly isolated, so that $\langle s_n \rangle$ cannot converge.

Two related but essentially different questions arise in the investigation of an infinite series: (1) Is it convergent or divergent? (2) If it is convergent, what is its sum? As a general rule, the second question is harder to answer than the first. One case in which both questions are easily answered is that of a geometric series.

Definition 13.5

A *geometric series* is an infinite series of the form

$$\sum_{n \in I} ar^{n-1} = a + ar + ar^2 + \cdots + ar^{n-1} + \cdots$$

where a and r are nonzero real numbers. The number r is called the *common ratio* of the geometric series.

Theorem 13.6

A geometric series $\sum_{n \in I} ar^{n-1}$ is convergent iff $|r| < 1$, in which case its sum is given by the formula $S = a/(1 - r)$.

Proof We show first that if $|r| < 1$, then the series converges to $a/(1 - r)$. For each positive integer n, the partial sum s_n is a geometric progression with first term a and common ratio r. By Theorem 2.15 we thus have

$$s_n = \frac{a(1 - r^n)}{1 - r} \qquad \text{for each } n \in I$$

and it follows from Prob. 12.19 that $\lim s_n = a/(1 - r)$. It remains to show that the geometric series is divergent for $|r| \geqslant 1$. But in this case we have for each $n \in I$,

$$|ar^n| = |a| \cdot |r|^n \geqslant |a|$$

Since $a \neq 0$, the series is divergent by the nth term test. □

We shall employ Theorem 13.6 as our principal tool in the discussion of sets of measure zero, as well as in our work with the Cantor set in Part Three. Meanwhile, we need one more basic result on series. Suppose we have an infinite series $\sum_{n \in I} x_n$ and a bijection f on I to I. Then f merely scrambles the set I of positive integers, mapping each positive integer n onto the positive integer $f(n)$. Hence the infinite series $\sum_{n \in I} x_{f(n)}$ must contain exactly the same terms as the series $\sum_{n \in I} x_n$ but arranged in a different order (unless f is the identity mapping on I). We call $\Sigma x_{f(n)}$ a rearrangement of the series Σx_n.

Definition 13.7

An infinite series $\sum_{n \in I} y_n$ is said to be a *rearrangement* of the infinite series $\sum_{n \in I} x_n$ iff there exists a bijection f on I to I such that $y_n = x_{f(n)}$ for each $n \in I$.

The question now arises of whether or not convergent series and their sums are preserved under rearrangements. Without imposing additional conditions on the original series, the answer is "no." A convergent series may have rearrangements which diverge, and it may have rearrangements which converge to different sums. However, it is convenient and sufficient for our purposes to restrict our attention to infinite series of *nonnegative* real numbers, and in this case the answer is "yes." We begin with the following fundamental convergence theorem for such series.

Theorem 13.8

An infinite series Σx_n of nonnegative real numbers is convergent iff its sequence $\langle s_n \rangle$ of partial sums is bounded.

Proof Since $x_n \geq 0$ for each n, the sequence $\langle s_n \rangle$ is nondecreasing, and the result follows from Theorem 12.15. □

Theorem 13.9

If Σx_n is an infinite series of nonnegative real numbers which converges to the real number S, then every rearrangement of Σx_n also converges to S.

Proof Let Σy_n be any rearrangement of Σx_n, and note that Σy_n is also a series of nonnegative real numbers. Then there is a bijection f on I to I such that $y_n = x_{f(n)}$ for each $n \in I$. If we denote by s_n the nth partial sum of Σy_n, it follows that for each $n \in I$

$$s_n = y_1 + y_2 + \cdots + y_n = x_{f(1)} + x_{f(2)} + \cdots + x_{f(n)}$$

Letting $N = \sup \{f(1), f(2), \ldots, f(n)\}$, it is easy to see that

$$s_n \leq x_1 + x_2 + \cdots + x_N \leq S$$

Hence Σy_n is convergent by Theorem 13.8. Furthermore, if we let $T = \Sigma y_n$, it follows from Theorem 12.15 that $T \leq S$. Since Σx_n is also a rearrangement of the convergent series Σy_n, we may interchange the roles of the two series and repeat the above argument to show that $S \leq T$. □

We might think of Theorem 13.9 as a rather limited extension of the commutative law for addition from finite sums to certain countably infinite sums of nonnegative real numbers. Suppose, for example, that $\{G_n\}$ is a countable collection of open intervals and that for each $n \in I$, the positive real number l_n is the length of the open interval G_n. If the infinite series Σl_n converges to a real number L, then we may call L the sum of the lengths of the open intervals in the collection $\{G_n\}$. By virtue of Theorem 13.9, the order in which the lengths are summed is immaterial.

Definition 13.10

A collection of sets $\{G_\alpha \mid \alpha \in H\}$ is said to *cover* a set K (or is said to be a *covering* of K) iff $K \subset \bigcup_{\alpha \in H} G_\alpha$. A subcollection of $\{G_\alpha\}$ which covers K is called a *subcover* (or a *subcovering*) of K.

Definition 13.11

A set K is said to be *of measure zero* iff given any real number $\epsilon > 0$ there is a countable collection $\{G_i\}$ of open intervals covering K such that the sum of the lengths of the open intervals in the collection $\{G_i\}$ is less than ϵ.

Theorem 13.12

Every countable set is of measure zero.

Proof Let K be a countable set and let $\epsilon > 0$ be given. If K is empty, it is clearly of measure zero. If K is nonempty, then there is a mapping $f: I \xrightarrow{\text{onto}} K$, so that we may write $K = \{k_i\}$, where $k_i = f(i)$ for each $i \in I$. If for each $i \in I$ we define the open interval G_i by

$$G_i = \left(k_i - \frac{\epsilon}{2^{i+2}},\ k_i + \frac{\epsilon}{2^{i+2}} \right)$$

then it is clear that $\{G_i\}$ is a countable collection of open intervals covering K. Moreover, for each i, the length of the open interval G_i is

$$k_i + \frac{\epsilon}{2^{i+2}} - \left(k_i - \frac{\epsilon}{2^{i+2}} \right) = \frac{2\epsilon}{2^{i+2}} = \frac{\epsilon}{2^{i+1}}$$

If we let L denote the sum of the lengths of the open intervals G_i for $i = 1, 2, \ldots$, it follows that

$$L = \frac{\epsilon}{2^2} + \frac{\epsilon}{2^3} + \frac{\epsilon}{2^4} + \cdots$$

$$= \frac{\epsilon}{2} \left(\frac{1}{2} + \frac{1}{2^2} + \frac{1}{2^3} + \cdots \right)$$

The expression in parentheses is a geometric series with $a = \frac{1}{2}$ and $r = \frac{1}{2}$. By Theorem 13.6, this geometric series converges to 1. Thus $L = \epsilon/2 < \epsilon$, and so K is a set of measure zero. □

Theorem 13.13

If $a < b$, the closed interval $[a,b]$ is not a set of measure zero.

Proof Let Δ be any countable collection of open intervals covering $[a,b]$. By the Heine-Borel theorem (Theorem 6.11), there is a finite subcollection $G_1, G_2, \ldots, G_n$ of Δ which covers $[a,b]$. We may assume that this finite collection is minimal in the sense that if any G_k, $1 \leq k \leq n$, is discarded, the remaining sets no longer cover $[a,b]$. For each $i = 1, 2, \ldots, n$, let $G_i = (a_i,b_i)$, and let us choose our notation so that $a_1 \leq a_2 \leq a_3 \leq \cdots \leq a_n$. Then it follows that

$$a_1 < a < b_1, a_2 < b_1 < b_2, a_3 < b_2 < b_3, \ldots, a_n < b < b_n$$

Denoting by S the sum of the lengths of the intervals, we thus have

$$\begin{aligned}
S &= (b_1 - a_1) + (b_2 - a_2) + (b_3 - a_3) + \cdots + (b_n - a_n) \\
&> (b_1 - a_1) + (b_2 - b_1) + (b_3 - b_2) + \cdots + (b_n - b_{n-1}) \\
&= b_n - a_1 \\
&> b - a \qquad \square
\end{aligned}$$

By virtue of Theorem 13.13 and the contrapositive of Theorem 13.12, we have now determined some uncountable sets.

Corollary 13.14

Every nondegenerate closed interval in R_1 is uncountable.

Corollary 13.15

Every open interval in R_1 is uncountable.

We shall see below that every finite set is countable, so that an uncountable set is necessarily infinite. In order to distinguish between infinite sets which are countable and those which are uncountable, we refer to the former sets as *countably infinite* or *denumerable* and to the latter sets as *uncountably infinite* or *non-denumerable*. Thus an infinite set is either denumerable or non-denumerable, and a countable set is either finite or denumerable.

Theorem 13.16

A nonempty set H is finite iff there exists a positive integer n such that $H = \{a_1, a_2, \ldots, a_n\}$.

Proof Half the theorem is already proved (see Theorem 4.11). For the converse, suppose there is a positive integer n such that $H = \{a_1, a_2, \ldots, a_n\}$. We want to show that H is finite, so let us assume by way of contradiction that H is infinite. By Prob. 12.13, there is a one-to-one mapping $f : I \to H$. Recalling that $I_n = \{1, 2, \ldots, n\}$, we define a mapping $g : H \to I_n$ by $g(a_i) = i$ for each $i = 1, 2, \ldots, n$. Clearly, the mapping g is one-to-one (in fact, g is a bijection on H to I_n), and hence $(g \circ f) : I \to I_n$ is one-to-one. But this contradicts the fact that I_n is finite (see Prob. 4.3).   □

Corollary 13.17

Every finite set is countable.

Proof Since the empty set is countable, we may suppose that H is a nonempty finite set. Then there is a positive integer n such that $H = \{a_1, a_2, \ldots, a_n\}$, and the mapping g^{-1} in the proof of Theorem 13.16 is a bijection on I_n to H. Since I_n is a retract of I, there is a mapping $f : I \xrightarrow{\text{onto}} I_n$, and hence $g^{-1} \circ f$ maps I onto H.   □

We close this chapter with an important characterization of infinite sets.

Theorem 13.18

A set H is infinite iff there is a bijection on H to a proper subset of H.

Proof Suppose that H is infinite. Then there is a one-to-one mapping $f : I \to H$. It is clear $\{f\} \subset H$ and that a point $x \in H$ is a member of $\{f\}$ iff $x = f(n)$ for some $n \in I$. Hence the set $H_1 = H - \{f(1)\}$ is a proper subset of H. We define the mapping g by

$$g(x) = \begin{cases} x & \text{if } x \in H - \{f\} \\ f(n+1) & \text{if } x = f(n) \end{cases}$$

and leave as an exercise the verification that g is a bijection on H to H_1.

For the converse, suppose that H is finite and let f be a bijection on H to K, where K is a subset of H. We shall show that K cannot be a proper subset of H. Let us assume by way of contradiction that K is a proper subset of H. Since f is one-to-one, it follows from Prob. 7.15 that $f(K)$ is a proper subset of $f(H) = K$. Define $K_1 = f(K)$. Since K_1 is a proper subset of K, another application of Prob. 7.15 asserts that $f(K_1)$ is a proper subset of

$f(K) = K_1$. Define $K_2 = f(K_1)$. Clearly we may continue this process to define inductively a sequence $\langle K_n \rangle$ of distinct proper subsets of H. But this is impossible since the finite set $H = \{a_1, a_2, \ldots, a_n\}$ for some positive integer n and hence has only 2^n subsets by Prob. 2.13. $\square$

PROBLEMS

13.1 The purpose of this problem is to prove that the *alternating harmonic series*

$$\sum_{n \in I} (-1)^{n+1} \frac{1}{n} = 1 - \frac{1}{2} + \frac{1}{3} - \frac{1}{4} + \cdots + \frac{1}{2n-1} - \frac{1}{2n} + \frac{1}{2n+1} - \cdots$$

converges. We call a partial sum s_n *even* or *odd* according as n is even or odd. Thus $\langle s_{2n} \rangle$ is the subsequence of even partial sums of $\langle s_n \rangle$, and $\langle s_{2n-1} \rangle$ is the subsequence of odd partial sums of $\langle s_n \rangle$.

(a) Prove that $\langle s_{2n} \rangle$ is an increasing sequence. *Hint:* Write

$$s_{2n} = s_{2n-2} + \left(\frac{1}{2n-1} - \frac{1}{2n} \right)$$

(b) Prove that $\langle s_{2n-1} \rangle$ is a decreasing sequence.

(c) Prove that every even partial sum is less than every odd partial sum.

(d) Conclude that $\langle s_{2n} \rangle$ converges to a real number S, and $\langle s_{2n-1} \rangle$ converges to a real number T, where $S \leqslant T$.

(e) Prove that $S = T$ and hence that the series converges.

Although we have established the convergence of this series, its sum is by no means apparent. It should be evident, however, that $\frac{1}{2} \leqslant S \leqslant 1$. Actually, $S = \ln 2$, where ln stands for logarithm to the base e.

13.2 Show that every subset of a set of measure zero is a set of measure zero.

13.3 If A and B are sets of measure zero, prove that $A \cup B$ and $A \cap B$ are sets of measure zero.

13.4 Prove that the set of irrational numbers in the interval $[0,1]$ is not a set of measure zero.

13.5 Prove that the set of irrational numbers is uncountable.

13.6 Show that the mapping g defined in the proof of Theorem 13.18 is a bijection on H to H_1.

13.7 Let K be the open interval $(0,2)$, and let $\{G_n\}$ be a countable family of open intervals defined as follows: $G_n = (0, 2 - 1/n)$ for each $n \in I$.

(a) Prove that $\{G_n\}$ is a covering of K.

(b) Prove that no finite subcollection of $\{G_n\}$ is a covering of K.

(c) Explain why this result does not contradict the Heine-Borel theorem.

13.8 Let K be the closed ray $\{x \mid x \geqslant 0\}$, and let $\{G_n\}$ be a countable family of open intervals defined as follows: $G_n = (n - 2, n + 2)$ for each $n \in I$.

(a) Prove that $\{G_n\}$ is a covering of K.

(b) Prove that no finite subcollection of $\{G_n\}$ is a covering of K.

(c) Explain why this result does not contradict the Heine-Borel theorem.

part three

Real Analysis

14
Closed Sets; Open Sets;
Cantor Product Theorem;
A Characterization of Open
Subsets of R_1

Let us recall that a point p is a cluster point of a set A iff every deleted open interval about p contains a point of A. The set A' of all cluster points of A is called the derived set of A. We now assign a special name to the set $A \cup A'$.

Definition 14.1

For any set A, the *closure* of A is the set $\bar{A} = A \cup A'$.

Some of the properties of closure are stated in the following theorem, the proof of which is immediate from Definition 14.1 and Theorem 6.4.

Theorem 14.2

Let A and B be sets in R_1. Then:

(a) $$\bar{\varnothing} = \varnothing$$

(b) $$\bar{R_1} = R_1$$

(c) $$A \subset \bar{A}$$
(d) $$A' \subset \bar{A}$$
(e) $$\text{If } A \subset B, \text{ then } \bar{A} \subset \bar{B}$$
(f) $$\overline{A \cup B} = \bar{A} \cup \bar{B}$$
(g) $$\overline{A \cap B} \subset \bar{A} \cap \bar{B}$$
(h) $$\bar{\bar{A}} = \bar{A}$$

Definition   14.3

(a)   A set A is said to be *closed* iff A contains all its cluster points; that is, A is closed iff $A' \subset A$.

(b)   A set A is said to be *open* iff its complement $C(A)$ is closed.

Theorem   14.4

A set A is closed iff $A = \bar{A}$.

Proof   By Theorem 1.7b, $A \cup A' = A$ iff $A' \subset A$. $\square$

Theorem   14.5

The following subsets of R_1 are closed sets:

(a) $\varnothing$
(b) R_1
(c)   Every finite set
(d)   Every closed interval
(e)   Every closed ray
(f)   The derived set A' of any set A
(g)   The closure $\bar{A}$ of any set A

Proof   Assertions (a), (b), and (g) follow immediately from (a), (b), and (h) of Theorem 14.2.   For (c), we note that if A is a finite set, then $A' = \varnothing$ by Corollary 6.9, and hence $A' \subset A$. Part (f) follows from Theorem 6.4d.   Since (d) and (e) are similar, we prove (d), leaving (e) as an exercise.   Let A be a closed interval $[a,b]$, and suppose $p < a$.   Then $(p - 1, p,a) \cap [a,b] = \varnothing$, and so $p \notin A'$. Similarly, if $p > b$, then $(b, \ p, \ p + 1) \cap [a,b] = \varnothing$, and so $p \notin A'$.   Hence $A' \subset A$. $\square$

Theorem   14.6

The intersection of any collection of closed sets is a closed set.

Proof   Let H be an index set and for each $\alpha \in H$, let A_α be a closed set.   We shall show that the set $A = \underset{\alpha \in H}{\cap} A_\alpha$ is closed.   By

Theorem 11.9b, $A \subset A_\alpha$ for each $\alpha \in H$, and hence $\bar{A} \subset \bar{A}_\alpha = A_\alpha$ by Theorem 14.2e. Thus $\bar{A} \subset A$, and it follows from Theorem 14.2c that $\bar{A} = A$.   □

The following corollary is an interesting characterization of the closure of a set, and its easy proof is left as an exercise.

Corollary 14.7

The closure of a set A in R_1 is the smallest closed set containing A, in the sense that if F is any closed set containing A, then $\bar{A} \subset F$. In particular, the set $\bar{A}$ is the intersection of all closed sets containing A.

Theorem 14.8

The union of any finite collection of closed sets is a closed set.

The proof of Theorem 14.8 is an easy exercise using induction and Theorem 14.2f. We emphasize the necessity of the word "finite" in the hypothesis of this theorem by giving an example of an infinite collection of closed sets whose union is not closed.

Example 14.9

For each $n \in I$, let A_n be the closed interval $[0, 2 - 1/n]$. Then each A_n is a closed set, but the union $A = \underset{n \in I}{\cup} A_n$ is not a closed set. The reader may verify that $A = [0,2)$, so that $2 \in \bar{A} - A$.

Theorem 14.10

(a) Every nonempty closed set which is bounded below contains its infimum.
(b) Every nonempty closed set which is bounded above contains its supremum.
(c) Every nonempty closed and bounded set contains both its supremum and its infimum.

The proof is immediate from Theorem 6.2 and the definition of a closed set.

Theorem 14.11   The Cantor product theorem for R_1

Let $\langle A_n \rangle$ be a sequence of subsets of R_1 satisfying the following conditions:

(a) A_1 is bounded.

(b) A_n is closed for each $n \in I$.
(c) A_n is nonempty for each $n \in I$.
(d) $A_{n+1} \subset A_n$ for each $n \in I$.

Then the set $A = \bigcap_{n \in I} A_n$ is closed, bounded, and nonempty.

Proof It follows from the hypotheses of this theorem that each set A_n is closed, bounded, and nonempty. Thus the set A is closed by Theorem 14.6, and A is bounded since it is a subset of the bounded set A_1. It remains only to show that A is nonempty. For each $n \in I$, we define

$$p_n = \inf A_n \quad \text{and} \quad q_n = \sup A_n$$

Then $p_n \in A_n$ for each $n \in I$, by Theorem 14.10. We have thus defined a sequence $f = \langle p_n \rangle$ of points of R_1. The following properties of this sequence f are immediate consequences of (a) and (d).

(i) f is nondecreasing.
(ii) $\{f\} \subset A_1$; in particular, q_1 is an upper bound for f.

Inasmuch as $\langle p_n \rangle$ is a bounded monotone sequence, it must converge to a real number p, by Theorem 12.15. We shall now prove that $p \in A$. Let us assume by way of contradiction that $p \notin A$. Then there exists a positive integer m such that $p \notin A_m$, and since $A_m = \bar{A}_m$, it follows that $p \notin \bar{A}_m$. Thus p is not a cluster point of A_m, and so there exists a deleted open interval (a,p,b) which contains no point of the set A_m. Hence $(a,b) \cap A_m = \varnothing$, and it follows from (d) that $A_n \cap (a,b) = \varnothing$ for every $n \geq m$. In particular, $p_n \notin (a,b)$ for $n \geq m$, so that $\{f_n\} \cap (a,b) = \varnothing$ for $n \geq m$. But this contradicts the fact that $f = \langle p_n \rangle$ converges to p.   □

The proofs of the following corollaries are left as exercises.

Corollary 14.12

If for each $n \in I$ the point q_n is defined as in the proof of Theorem 14.11, then the sequence $\langle q_n \rangle$ is nonincreasing and converges to a point q in $A = \bigcap_{n \in I} A_n$.

Corollary 14.13

Let p be defined as in the proof of Theorem 14.11 and q as in Corollary 14.12. Then every point x in A satisfies $p \leqslant x \leqslant q$.

Corollary 14.14

The set A is a singleton iff $p = q$.

From Definition 14.3 we see that a set A in R_1 is not closed iff there is a point $p \in R_1 - A$ such that p is a cluster point of A. Now by Corollary 12.6, the point $p \in R_1 - A$ is a cluster point of A iff there is a sequence $\langle p_n \rangle$ of points of A converging to p. Clearly, the sequence $\langle p_n \rangle$ is a convergent sequence in R_1, and hence a Cauchy sequence in R_1. It follows at once that a set A in R_1 is not closed iff there is a Cauchy sequence of points of A which is not convergent in A. But this is exactly what we mean by the statement "A is not complete." Hence a subset A of R_1 is not closed iff A is not complete, and we state the more convenient contrapositive form of this result as our next theorem.

Theorem 14.15

A subset A of R_1 is complete iff A is closed.

Our next theorem is an extremely useful characterization of open sets in R_1.

Theorem 14.16

A set A in R_1 is open iff for each point $p \in A$ there exists an open interval (a,b) such that $p \in (a,b) \subset A$.

Proof Suppose that A is open and let p be any point of A. Then $C(A)$ is closed by Definition 14.3, and so $p \notin \overline{C(A)}$. Since p is not a cluster point of $C(A)$, there exists a deleted open interval (a,p,b) which contains no point of $C(A)$. Therefore $p \in (a,b) \subset A$.

For the converse, let p be any point of A, and suppose there exists an open interval (a,b) such that $p \in (a,b) \subset A$. Then p is not a cluster point of $C(A)$. Thus $C(A)$ contains all its cluster points and must therefore be closed. Hence A is open. $\square$

We wish to emphasize that as used in analysis the words "open" and "closed" are not antonyms. In particular, there are sets—such as $[a,b)$—which are neither open nor closed. Also, the empty set $\varnothing$ and R_1 are both open and closed. Thus we cannot conclude that a set is open merely because it is not closed or that it is closed merely because it is not open.

Since open sets are complements of closed sets, many properties of open sets can be easily established by working with the complementary closed sets in conjunction with De Morgan's laws. The proof of our next theorem is thus an easy consequence of Theorems 14.5, 14.6, and 14.8.

Theorem 14.17

The following subsets of R_1 are open sets:

(a) $\varnothing$
(b) R_1
(c) Every open interval
(d) Every open ray
(e) The union of any collection of open sets
(f) The intersection of any finite collection of open sets

The student is required in the problems to prove this theorem both by the technique indicated above and by a method which does not employ De Morgan's laws. It is also required to give an example showing that the word "finite" is essential in (f).

A glance at Theorems 14.5 and 14.17 shows that each of the sets R_1 and $\varnothing$ is both open and closed. It will follow from our next theorem that these are the only subsets of R_1 with this property.

Theorem 14.18

Let R_1 be the union of two disjoint nonempty sets H and K. Then each of the following statements is true:

(a) H and K are not both open.
(b) H and K are not both closed.
(c) H is not both open and closed.
(d) K is not both open and closed.

Proof We leave as an exercise the proof that these four statements are equivalent, and so we prove only (a). We are given that H and K are disjoint nonempty sets whose union is R_1, and let us assume by way of contradiction that H and K are both open. By Corollary 5.14, H and K cannot be separated; so we must have $H \cap \bar{K} \neq \varnothing$ or $\bar{H} \cap K \neq \varnothing$. Suppose $p \in H \cap \bar{K}$. Since H is open, there exists an open interval (a,b) such that $p \in (a,b) \subset H$. Then $(a,p,b) \cap K = \varnothing$, which contradicts the fact that $p \in \bar{K}$. A similar contradiction occurs if $p \in \bar{H} \cap K$.   □

Corollary 14.19

R_1 and $\varnothing$ are the only subsets of R_1 which are both open and closed.

Proof Let H be a subset of R_1 which is both open and closed. Then $C(H)$ is also open and closed. Since R_1 is the union of the disjoint sets H and $C(H)$, the contrapositive of Theorem 14.18 asserts that these sets cannot both be nonempty. Thus either $H = \varnothing$, or else $C(H) = \varnothing$, in which case $H = R_1$.   ☐

We conclude this section with a useful characterization of the structure of open sets in R_1. Our proof will use the following preliminary result.

Theorem   14.20

Every open interval (a,b) in R_1 contains a rational number.

Proof We lose no generality if we assume $a > 0$. Defining $d = b - a > 0$, the Archimedean property asserts the existence of a positive integer N such that $Nd > 1$ as well as a positive integer M such that $M(1/N) > b$. Let K be the set of positive integers k for which $k/N \geqslant b$. Then K is nonempty since $M \in K$, and so by the well-ordering property, K has a smallest element y. Thus $y/N \geqslant b$, and y is the smallest positive integer for which this inequality is true. Clearly $y > 1$, since $1/N < d < b$ and $y(1/N) > b$. Hence for the positive integer $y - 1$ we must have $(y - 1)/N < b$. It remains to show that $(y - 1)/N > a$. Using the above relations, we have

$$a = b - d \leqslant \frac{y}{N} - d < \frac{y}{N} - \frac{1}{N} = \frac{y-1}{N}$$

Therefore the rational number $(y - 1)/N$ is between a and b.   ☐

Theorem   14.21

A subset H of R_1 is open iff H is the union of a countable collection of open intervals.

Proof If H is the union of any collection of open intervals, then H is open by Theorem 14.17. For the converse, suppose H is a nonempty open subset of R_1. We want to show that H is the union of a countable collection of open intervals. Since the set Q of all rational numbers is countable, it follows that $H \cap Q$ is countable; so let $\langle r_i \rangle$ be the sequence of all rational numbers which are contained in H. Defining the subset J of $I \times I$ by

$$J = \{\langle i,j \rangle \mid \langle i,j \rangle \in I \times I;\ r_i < r_j;\ \text{and}\ r_i, r_j \in \langle r_i \rangle\}$$

we know that J is countable. Hence the set K defined by

$$K = \bigcup_{\langle i,j \rangle \in J} \{ (r_i, r_j) \mid (r_i, r_j) \subset H \}$$

is the union of a countable collection of open intervals. We shall complete the proof by showing that $K = H$. It is clear from the definition of K that $K \subset H$. For the reverse inclusion, suppose p is any point of H. Since H is open, there exists an open interval (a,b) such that $p \in (a,b) \subset H$. By Theorem 14.20, the open interval (a,p) contains a rational number q, and since we also have $q \in H$, it follows from the definition of the sequence $\langle r_i \rangle$ that $q = r_m$ for some positive integer m. Similarly, there is a rational number r_n of the sequence $\langle r_i \rangle$ such that $r_n \in (b,p)$. Clearly, $(r_m, r_n) \subset (a,b) \subset H$, and so $(r_m, r_n) \subset K$. Since $p \in (r_m, r_n)$, it follows that $p \in K$.   □

PROBLEMS

14.1 Complete the proof of Theorem 14.5.

14.2 Prove Corollary 14.7.

14.3 Prove Theorem 14.8.

14.4 Prove Corollary 14.12.

14.5 Prove Corollary 14.13.

14.6 Prove Corollary 14.14.

14.7 Show that if each set A_n in the Cantor product theorem is the closed interval $[p_n, q_n]$, their intersection is the closed interval $[p,q]$. Furthermore, if the lengths of the closed intervals tend to 0 as n increases, then $p = q$ and the intersection consists of a single point. (In this form, Theorem 14.11 is sometimes called the *nested intervals theorem*.)

14.8 Show that the Cantor product theorem remains true if condition (a) of the hypotheses is relaxed to read as follows:

(a') The set A_k is bounded for some $k \in I$.

14.9 For each $n \in I$, let A_n be the closed ray $\{ x \in R_1 \mid x \geq n \}$. Show that

$$A = \bigcap_{n \in I} A_n = \varnothing$$

Explain why this result does not contradict Theorem 14.11.

14.10 For each $n \in I$, let A_n be the open interval $(0, 1/n)$. Show that $A = \bigcap_{n \in I} A_n = \varnothing$. Explain why this result does not contradict Theorem 14.11.

14.11 Let z be a real number and H a nonempty subset of R_1.

$$d(z,H) = \inf \{ |z - x| \mid x \in H \}$$

is called the *distance from the point z to the set H*.

  (a) Prove that if H is closed and $d(z,H) = 0$, then $z \in H$.

  (b) Prove that a point z is in the closure of H iff $d(z,H) = 0$.

14.12 Prove that if H is a nonempty closed subset of R_1 and $z \in R_1 - H$, then there exists at least one point $x_o \in H$ such that $d(z,H) = |x_0 - z|$. *Hint:* For each $n \in I$, define $A_n = \{x \in H \mid |z - x| \leqslant d(z,H) + 1/n\}$, and apply the Cantor product theorem.

14.13 Let A be a connected set in R_1 and A^* any set such that $A \subset A^* \subset \bar{A}$. Prove that A^* is connected. This says in particular that the closure of a connected set is connected.

14.14 Prove Theorem 14.17 using the appropriate properties of closed sets in conjunction with De Morgan's laws.

14.15 Prove Theorem 14.17 without using De Morgan's laws.

14.16 Prove that the word "finite" is essential in the statement of Theorem 14.17f by giving an example of an infinite collection of open sets whose intersection is not open.

14.17 Prove that the four assertions (a), (b), (c), and (d) of Theorem 14.18 are equivalent.

14.18 Show that every open interval in R_1 contains an infinite number of distinct rationals.

14.19 Let Q be the set of all rational numbers. Show that $\bar{Q} = R_1$, and conclude that every irrational number is the limit of a sequence of rational numbers.

14.20 Let L_1 and L_2 be disjoint nonempty sets whose union is R_1. Suppose that for any pair of points x and y such that $x \in L_1$ and $y \in L_2$, we have $x < y$. Prove that either L_1 has a greatest element or L_2 has a least element. (This is usually called the *Dedekind cut property*.)

14.21 Define the following two subsets of R_1:

$$L_1 = \{x \mid x \geqslant 0 \text{ and } x^2 < 2\} \cup \{x \mid x < 0\}$$
$$L_2 = \{x \mid x > 0 \text{ and } x^2 > 2\}$$

Prove that each of the sets L_1 and L_2 is nonempty and open. What conclusion can you draw from these facts? Do not use the square root of 2 anywhere in your proof.

15

Compactness; Further Results on Continuous Functions

In Definition 13.10, we defined a covering of a set K as a collection of sets $\{G_\alpha \mid \alpha \in H\}$ such that $K \subset \bigcup_{\alpha \in H} G_\alpha$, where H is an index set. A covering is called *finite* or *countable* according as the index set H is finite or countable, respectively, and is called *open* if G_α is an open set for each $\alpha \in H$. We shall now use these concepts to define three types of compactness and then prove that in R_1 all three are equivalent.

Definition 15.1

A subset K of R_1 is said to be:

(a) *Compact* iff every open covering of K contains a finite open subcovering of K.

(b) *Countably compact* iff every countable open covering of K contains a finite open subcovering of K.

(c) *Sequentially compact* iff every sequence of points of K has a subsequence which is convergent in K.

The negation of (c) reads as follows: A subset K of R_1 is not sequentially compact iff there exists a sequence f of points of K such that no subsequence of f is convergent in K. In the proof of our major theorem, we shall use a form of this negation which appears stronger but which is actually equivalent by virtue of the following theorem, the proof of which is left as an exercise.

Theorem 15.2

A subset K of R_1 is not sequentially compact iff there exists a sequence f of distinct points of K such that no subsequence of f is convergent in K.

Theorem 15.3

The following four properties of a subset K of R_1 are equivalent:

(a) K is closed and bounded.
(b) K is compact.
(c) K is countably compact.
(d) K is sequentially compact.

Proof Since all four conditions are trivially satisfied if K is empty or a singleton, we may suppose that K contains at least two points.

(a) implies (b) We are given that K is a bounded closed set containing at least two points, and we must show that K is compact. Let $\{G_\alpha \mid \alpha \in H\}$ be any open covering of K. Letting $a = \inf K$ and $b = \sup K$, we know that $a < b, a \in K, b \in K$, and $K \subset [a,b]$. Defining $G_0 = (a,b) \cap C(K)$, it follows that G_0 is an open set, and hence $\{G_\alpha\} \cup G_0$ is an open covering of $[a,b]$. By Theorem 14.21, each open set in this covering is the union of a countable collection of open intervals. We thus define Δ as the collection of all open intervals A_β such that $A_\beta \subset G_\alpha$ for some $\alpha \in H$, or $A_\beta \subset G_0$. Then Δ is a covering of $[a,b]$, so that by Theorem 6.11 there is a finite subcollection $A_{\beta_1}, A_{\beta_2}, \ldots, A_{\beta_n}$ of open intervals in Δ which is a covering of $[a,b]$. Since $K \subset [a,b]$, this finite collection of open intervals is also a covering of K. Now suppose that $A_{\beta_i} \subset G_0$ for some i such that $1 \leq i \leq n$. Then $A_{\beta_i} \cap K = \varnothing$; so we may discard A_{β_i}, and the remaining open intervals still cover K. To simplify notation, we may thus suppose that $A_{\beta_1}, A_{\beta_2}, \ldots, A_{\beta_n}$ cover K and that for each $i = 1, 2, \ldots, n$, $A_{\beta_i} \subset G_{\alpha_i}$ for some $\alpha_i \in H$. Then $G_{\alpha_1}, G_{\alpha_2}, \ldots, G_{\alpha_n}$ is a finite subcollection of $\{G_\alpha\}$ which covers K, and so K is compact.

(b) *implies* (c) This is immediate from the definitions of compact and countably compact.

(c) *implies* (d) We shall prove the contrapositive, so suppose that K is not sequentially compact. By Theorem 15.2, there exists a sequence $f = \langle p_n \rangle$ of distinct points of K such that no subsequence of f is convergent in K. Defining $H = \{f\}$, the contrapositive of Prob. 12.20 asserts that no point of K is a cluster point of H. In particular, we must then have

(i) No point of H is a cluster point of H, and
(ii) No point of $K - H$ is a cluster point of H.

It follows from (i) that if $h = p_n$ is any point of H, then there exists a deleted open interval (a,h,b) such that $(a,h,b) \cap H = \varnothing$, and hence $(a,b) \cap H = \{h\}$. Now for each $n \in I$, define $\mathcal{M}_n$ as the collection of all open sets G_α such that $G_\alpha \cap H = \{p_n\}$. Then each $\mathcal{M}_n$ is nonempty, so that by the axiom of choice for sequences, there exists a sequence $\langle G_n \rangle$ of open sets such that $G_n \in \mathcal{M}_n$ for each $n \in I$. Since $f = \langle p_n \rangle$ is a sequence of distinct points, it is evident that $\{G_n\}$ is a countable collection of open sets covering H, and each point of H is in exactly one set of the covering.

It follows from (ii) that $K - H \subset C(\bar{H})$, and so the open set $C(\bar{H})$ is an open covering of $K - H$. Clearly then, $\{G_n\} \cup C(\bar{H})$ is a countable open covering of K which has no finite open subcovering of K. Therefore K is not countably compact.

(d) *implies* (a) We must show that if a subset K of R_1 is sequentially compact, then K is closed and K is bounded. We shall prove the contrapositive, which asserts that if K is not closed or if K is not bounded, then K is not sequentially compact.

Suppose K is not closed. Then there exists a point $p \in K'$ such that $p \notin K$. By Corollary 12.6, there exists a strictly monotone sequence $f = \langle p_n \rangle$ of points of K converging to p. Since $p \notin K$, it is clear that no subsequence of f is convergent in K. Therefore K is not sequentially compact.

Suppose K is not bounded. By Theorem 12.2, there is a sequence f of points of K which is monotone and uniformly isolated. Since every subsequence of f is uniformly isolated, it follows that no subsequence of f is convergent in K. Therefore K is not sequentially compact. $\square$

The equivalence of (a) and (b) of Theorem 15.3 is a considerably stronger result than the statement of Theorem 6.11 and is the usual form for expressing the Heine-Borel theorem in R_1. Also, the equivalence of (a) and (d) is yet another form of the important Bolzano-Weierstrass theorem. To emphasize their importance, we list them as corollaries and attach the appropriate names.

Corollary 15.4 The Heine-Borel theorem

A subset K of R_1 is compact iff K is closed and bounded.

Corollary 15.5 Bolzano-Weierstrass theorem for sequential compactness

A subset K of R_1 is sequentially compact iff K is closed and bounded.

We shall presently show that compactness is preserved by continuous functions, but first it is necessary to reformulate the definition of continuity. Suppose that $f:A \rightarrow R_1$ and $p \in A$. Then we know that f is continuous at p iff given any $\epsilon > 0$, there exists $\delta > 0$ such that $f((p - \delta, p + \delta) \cap A) \subset (f(p) - \epsilon, f(p) + \epsilon)$. In other words, f is continuous at p iff given any open interval G centered at $f(p)$, there exists an open interval centered at p whose image under f is contained in G. This property can be better stated in terms of inverse images.

Definition 15.6

Let $f:A \rightarrow R_1$ be a mapping. For any subset B of R_1, we denote by $f^{-1}(B)$ the *inverse image* of the set B under the mapping f, where

$$f^{-1}(B) = \{x \in A \,|\, f(x) \in B\}$$

In particular $f^{-1}(B) = f^{-1}[B \cap f(A)]$, since $B \cap f(A) = \varnothing$ iff $f^{-1}(B) = \varnothing$.

A word of caution is in order to avoid confusion between the similar notations for inverse images and inverse functions. Note that the inverse image $f^{-1}(B)$ of a set B is a well-defined set for any mapping f, regardless of whether or not f^{-1} exists. In the event that f^{-1} does exist, then the inverse image $f^{-1}(B)$ coincides with the image of B under the mapping f^{-1}.

The following theorem should now be self-evident.

Theorem 15.7

A function $f:A \rightarrow R_1$ is continuous at a point $p \in A$ iff given any open interval (a,b) centered at $f(p)$ there exists an open interval (c,d) centered at p such that

$$f^{-1}((a,b)) \supset (c,d) \cap A$$

This "local" criterion for continuity of f at a point p of its domain may now be used to derive a "global" criterion for continuity of f on its entire domain.

Theorem 15.8 Open-set criterion for continuity

A function $f:A \to R_1$ is continuous iff given any open set K in R_1 there exists an open set G such that $f^{-1}(K) = G \cap A$.

Proof Suppose f is continuous and for each point p of $f^{-1}(K)$, let (a_p, b_p) be an open interval centered at $f(p)$ and contained in K. By Theorem 15.7, there exists an open interval (c_p, d_p) centered at p such that $f^{-1}((a_p, b_p)) \supset (c_p, d_p) \cap A$. Define

$$G = \bigcup_{p \in f^{-1}(K)} (c_p, d_p)$$

The simple proof that $f^{-1}(K) = G \cap A$ is left as an exercise.
For the converse, let p be any point of A, and let $\epsilon > 0$ be given. Define $K = (f(p) - \epsilon, \; f(p) + \epsilon)$. There exists $\delta > 0$ such that $(p - \delta, \; p + \delta) \subset G$, where G is the open set such that $f^{-1}(K) = G \cap A$. Thus f is continuous at p by the definition of continuity.   □

The open-set criterion for continuity is indispensable in both analysis and topology. As an illustration, we use this criterion to prove that compactness is preserved under continuous mappings.

Theorem 15.9

If H is a compact set in the domain of a continuous mapping f, then $f(H)$ is compact.

Proof Let $\{K_\alpha\}$ be any open covering of $f(H)$. By Theorem 15.8, there exists for each α an open set G_α such that $f^{-1}(K_\alpha) = G_\alpha \cap H$. Clearly, $\{G_\alpha\}$ is an open covering of the compact set H. Thus there exists a finite open subcovering $\{G_1, G_2, \ldots, G_n\}$ of H. Hence $\{K_1, K_2, \ldots, K_n\}$ is a finite open subcovering of $f(H)$.   □

Several corollaries to Theorem 15.9 are immediate consequences of Theorem 15.3. Some of them are listed below to emphasize their importance.

Corollary 15.10

Continuous mappings preserve sequential compactness; i.e., if H is a sequentially compact set in the domain of a continuous mapping f, then $f(H)$ is sequentially compact.

Corollary 15.11

If H is a closed and bounded set in the domain of a continuous mapping f, then $f(H)$ is closed and bounded.

   Although sets which are both closed and bounded are preserved under continuous mappings, we remark that closed sets are not necessarily preserved under continuous mappings, nor are bounded sets. Examples are requested in the problems.
   Because a special case of Corollary 15.11 has wide application in the calculus, we list it as a matter of general interest.

Corollary 15.12

A continuous function f on a closed interval $[a,b]$ is bounded and attains both a maximum value and a minimum value on the interval. More specifically, there are points x_1 and x_2 in $[a,b]$ such that

$$f(x_1) \leqslant f(x) \leqslant f(x_2) \qquad \text{for } every \text{ } x \text{ in } [a,b]$$

Theorem 15.13 Closed-set criterion for continuity

A function $f:A \to R_1$ is continuous iff given any closed set H in R_1 there exists a closed set F such that $f^{-1}(H) = F \cap A$.

Proof Suppose f is continuous, and let H be any closed subset of R_1. Define $K = R_1 - H$. Then K is open, so that by Theorem 15.8 there exists an open set G such that $f^{-1}(K) = G \cap A$. Now define $F = R_1 - G$, noting that F is closed and that

$$\begin{aligned} f^{-1}(H) &= f^{-1}(f(A) \cap H) = f^{-1}(f(A) - K) \\ &= A - f^{-1}(K) = A - (G \cap A) \\ &= A \cap F \end{aligned}$$

   For the converse we also use Theorem 15.8; so let K be any open subset of R_1. Define $H = R_1 - K$. Then H is closed, and so by hypothesis there exists a closed set F such that $f^{-1}(H) = F \cap A$. Now define $G = R_1 - F$, noting that G is open and that

$$f^{-1}(K) = f^{-1}(f(A) \cap K) = f^{-1}(f(A) - H)$$
$$= A - f^{-1}(H) = A - (F \cap A)$$
$$= A \cap G$$

Therefore f is continuous by Theorem 15.8.   □

Another useful "local" criterion for continuity involves the preservation of convergent sequences.

Theorem 15.14   Sequential criterion for continuity

A function $f:A \rightarrow R_1$ is continuous at a point $p \in A$ iff given any sequence $\langle p_n \rangle$ of points of A converging to p, the sequence $\langle f(p_n) \rangle$ converges to $f(p)$.

Proof   Suppose f is continuous at p, and let $\langle p_n \rangle$ be any sequence of points of A such that $\lim p_n = p$. We must show that $\lim f(p_n) = f(p)$; so let $\epsilon > 0$ be given. By Theorem 15.7, there exists $\delta > 0$ such that $f^{-1}((f(p) - \epsilon, f(p) + \epsilon)) \supset (p - \delta, p + \delta) \cap A$. Since $\lim p_n = p$, there exists a positive integer N such that $|p_n - p| < \delta$ for all $n \geqslant N$. It follows that $|f(p_n) - f(p)| < \epsilon$ for all $n \geqslant N$.

For the converse we prove the contrapositive; so suppose that f is not continuous at p. By Theorem 15.7 there exists an $\epsilon > 0$ such that for any $\delta > 0$, $f^{-1}((f(p) - \epsilon, f(p) + \epsilon))$ does not contain $(p - \delta, p + \delta) \cap A$. For each $n \in I$ define the set

$$K_n = \left[\left(p - \frac{1}{n}, \, p + \frac{1}{n} \right) \cap A \right] - f^{-1}((f(p) - \epsilon, f(p) + \epsilon))$$

Then K_n is nonempty for each $n \in I$, so that by the axiom of choice for sequences, there exists a sequence $\langle p_n \rangle$ of points such that $p_n \in K_n$ for each $n \in I$. Clearly $\langle p_n \rangle$ is a sequence of points of A such that for each $n \in I$, $p_n \in (p - 1/n, \, p + 1/n)$ but $f(p_n) \notin (f(p) - \epsilon, f(p) + \epsilon)$; that is, $|p_n - p| < 1/n$, and $|f(p_n) - f(p)| \geqslant \epsilon$. We thus have $\lim p_n = p$, but $\langle f(p_n) \rangle$ does not converge to $f(p)$.   □

Theorem 15.15

Connected sets are preserved by continuous mappings; i.e., if H is a connected set in the domain of a continuous mapping f, then $f(H)$ is connected.

Proof   We prove the contrapositive; so suppose H is a subset of the domain of a continuous mapping f and that $f(H)$ is not connected. Then by Theorem 5.12, $f(H)$ is the union of two separated sets A and B; that is, A and B are disjoint nonempty sets whose union is

$f(H)$, and neither set contains a cluster point of the other. It is easy to verify that $f^{-1}(A)$ and $f^{-1}(B)$ are disjoint nonempty sets whose union is H. Furthermore, neither contains a cluster point of the other. To see this, assume that $f^{-1}(A)$ contains a point p which is a cluster point of $f^{-1}(B)$. Then there is a sequence $\langle p_n \rangle$ of distinct points of $f^{-1}(B)$ such that $\lim p_n = p$. By Theorem 15.14, we then have $\lim f(p_n) = f(p)$. It follows that $f(p) \in A \cap \bar{B}$, contradicting the fact that A and B are separated. Hence H is the union of the separated sets $f^{-1}(A)$ and $f^{-1}(B)$, so that, by Theorem 5.12, H is not connected.   □

Corollary 15.16   Intermediate value theorem

Let f be continuous on a connected subset H of R_1, and let $x_1 < x_2$ be points of H such that $f(x_1) \neq f(x_2)$. Then for any real number c between $f(x_1)$ and $f(x_2)$ there exists a point x in H such that $x_1 < x < x_2$ and $f(x) = c$.

The intermediate value theorem is often stated in the calculus as follows: "A function continuous on an interval assumes as a value (at least once) every number between any two of its values." We first encountered this important theorem in algebra, when we used the fact that if f is a polynomial such that $f(a)$ and $f(b)$ have opposite signs, then the equation $f(x) = 0$ has a real root between a and b.

We close this chapter with an interesting application of the intermediate value theorem which will be useful in Chap. 16. This result is generally known as the universal chord theorem and is discussed by R. P. Boas, Jr., in Volume 13 of the Carus Mathematical Monographs, "A Primer of Real Functions," published by The Mathematical Association of America.

Theorem 15.17   Universal chord theorem

Let f be continuous on the closed interval $[a,b]$, and suppose that $f(a) = f(b)$. Then there are real numbers c and d such that

(a) $$a < c < d < b$$

(b) $$d - c < \frac{b-a}{2}$$

(c) $$f(c) = f(d)$$

Proof Let $h = (b - a)/3$, and define g on $[a, a + 2h]$ by

$$g(x) = f(x + h) - f(x) \qquad \text{for } x \in [a, a + 2h]$$

Note that g is continuous on $[a, a + 2h]$ and that

$$
\begin{aligned}
g(a) + g(a + h) + g(a + 2h) &= f(a + h) - f(a) \\
&\quad + f(a + 2h) - f(a + h) \\
&\quad + f(a + 3h) - f(a + 2h) \\
&= f(a + 3h) - f(a) \\
&= f(b) - f(a) \\
&= 0
\end{aligned}
$$

It follows at once that either all three of the real numbers $g(a)$, $g(a + h)$, $g(a + 2h)$ must be zero or two of them must have opposite signs. If all three are zero, then in particular, $0 = g(a + h) = f(a + 2h) - f(a + h)$, and the points $c = a + h$, $d = a + 2h$ clearly satisfy the requirements of the theorem. If two of them have opposite signs, then by the intermediate value theorem there is a point p, necessarily in $(a, a + 2h)$, such that $g(p) = 0$. Hence $f(p) = f(p + h)$, and the points $c = p$, $d = p + h$ satisfy the requirements of the theorem.   □

Corollary 15.18

Let f be continuous on the closed interval $[a,b]$, and suppose that $f(a) = f(b)$. Then there are two sequences $\langle a_n \rangle$ and $\langle b_n \rangle$ of real numbers such that for each $n \in I$

(a) $a < a_n < a_{n+1} < b_{n+1} < b_n < b$

(b) $b_n - a_n < \dfrac{b - a}{2^n}$

(c) $f(a_n) = f(b_n)$

PROBLEMS

15.1   Prove Theorem 15.2. (For the nontrivial part, use Theorem 11.13.)

15.2   Let K be the open interval $(-1,1)$. Show directly from the definition, i.e., without using Theorem 15.3 that:
   (a) K is not compact.
   (b) K is not sequentially compact.

15.3   Let $K = (-1,1)$, and for each $\alpha \in K$ define $G_\alpha = (\alpha - \frac{1}{4}, \alpha + \frac{1}{4})$. Verify that $\{G_\alpha\}$ is an open covering of K, and then determine a finite open subcovering of K. Does this result contradict the Heine-Borel theorem?

15.4   Let $K = (-1,1)$, and define a sequence $f = \langle p_n \rangle$ of points of K as follows:

$$
p_n = \begin{cases} 1 - \dfrac{1}{n} & \text{for } n \text{ odd} \\[2mm] -\dfrac{1}{n} & \text{for } n \text{ even} \end{cases}
$$

(a) Show that f is not convergent in K.

(b) Show that f has a subsequence which is convergent in K.

(c) Does the result of (b) contradict the sequential form of the Bolzano-Weierstrass theorem?

15.5 Let $f:R_1 \to R_1$ be defined by $f(x) = x^2 + 1$ for every x in R_1. For $B_1 = [1,5]$ and $B_2 = [-10,5]$, show that $f^{-1}(B_1) = f^{-1}(B_2) = [-2,2]$.

15.6 Complete the first part of the proof of Theorem 15.8 by showing that $f^{-1}(K) = G \cap A$.

15.7 Give an example of a continuous function f on a closed set H such that $f(H)$ is not closed.

15.8 Give an example of a continuous function f on a bounded set H such that $f(H)$ is not bounded.

15.9 Give an example of a function which is defined on $[0,1]$ but which is not bounded.

15.10 Give an example of a function which is bounded on the closed interval $[0,1]$ but which attains neither a maximum value nor a minimum value on the interval.

15.11 Give an example of a continuous bijection on $(0,1)$ to R_1.

15.12 Show that the equation $x^6 + 6x - 5 = 0$ has a real root between 0 and 1.

15.13 Let $f:A \to R_1$ be a mapping, and let H and K be subsets of R_1. Prove that:

(a) $$f^{-1}[f(A)] = A$$

(b) $$f^{-1}(H \cup K) = f^{-1}(H) \cup f^{-1}(K)$$

(c) $$f^{-1}(H \cap K) = f^{-1}(H) \cap f^{-1}(K)$$

(d) $$f[f^{-1}(H)] = H \cap f(A)$$

(e) $$f^{-1}[f(B)] \supset B \quad \text{for any subset } B \subset A$$

15.14 Give an example of a mapping $f:A \to R_1$ and a subset $B \subset A$ such that $f^{-1}[f(B)] \ne B$. Can you determine an "iff" condition on f to ensure that $f^{-1}[f(B)] = B$ for every $B \subset A$?

15.15 Let $J = [0,1]$, and let $f:J \to J$ be any continuous mapping. Prove that there exists a point $x_0 \in J$ such that $f(x_0) = x_0$. *Hint:* Consider the mapping $g:R_1 \to R_1$ defined by $g(x) = f(x) - x$ for $x \in R_1$.

15.16 Let H be the union of two separated sets A and B, and let $f:H \to R_1$ be a function defined by $f(x) = 0$ if $x \in A$, $f(x) = 1$ if $x \in B$. Use the open-set criterion to prove that f is continuous.

15.17 Let $f:H \to R_1$ be a continuous function such that $f(H) = \{0\} \cup \{1\}$. Define $A = f^{-1}(\{0\})$ and $B = f^{-1}(\{1\})$. Prove that A and B are separated and that $H = A \cup B$.

15.18 Prove that a set H is separated iff there exists a continuous function mapping H onto $\{0\} \cup \{1\}$.

15.19 Give an alternate proof of Theorem 15.15 using the results of Prob. 15.18.

15.20 Prove that if $f:A \to R_1$ is continuous at the point $p \in A$ and if $f(p) \ne 0$, then there exists a real number $\delta > 0$ such that $f(x) \ne 0$ for $x \in (p - \delta, p + \delta) \cap A$.

15.21 Let f be continuous on the closed interval $[a,b]$, and suppose that $f(a) = f(b)$. Prove that there exists a sequence $\langle [a_n,b_n] \rangle$ of nested closed intervals such that

$f(a_n) = f(b_n)$ for each $n \in I$ and $\bigcap\limits_{n \in I} [a_n, b_n] = \{p\}$, where $p \in (a,b)$. *Hint:* Use Corollary 15.18 and the nested intervals theorem.

15.22   Let $\langle [a_n, b_n] \rangle$ be a sequence of nested closed intervals such that

$$\bigcap\limits_{n \in I} [a_n, b_n] = \{p\}$$

where $p \in (a_1, b_1)$, and let g be a continuous function on $[a_1, b_1]$ such that $g(a_n)$ and $g(b_n)$ have opposite signs for each $n \geq N$.   Prove that $g(p) = 0$.

16
Differentiable Functions; Properties of the Derivative

Recall that for any function $f:A \to R_1$ there is associated with each point $p \in A$ a function $Q_{f,p}:A - \{p\} \to R_1$ called the *difference quotient* of f at p (see Definition 8.7) defined by

$$Q_{f,p}(x) = \frac{f(x) - f(p)}{x - p} \qquad \text{for each } x \in A - \{p\}$$

The usual procedure in calculus is to define the derivative of f at p to be the limit of $Q_{f,p}(x)$ as x approaches p, provided the limit exists. This procedure necessitates an extensive preliminary discussion of limits and a maze of ϵ-δ proofs which seem to baffle many students. Accordingly, we have chosen to treat differentiability as a simple extension problem involving continuity, a procedure which appears in "Fundamentals of Abstract Analysis," by Andrew M. Gleason (Addison-Wesley, 1966).

To illustrate the extension problem, suppose that $f:A \to R_1$ is a given function and that $p \in A \cap A'$. (We require $p \in A'$ since we do not want f to be differentiable at an isolated point of its domain.) Now, can we extend the function $Q_{f,p}:A - \{p\} \to R_1$ to a (necessarily

unique) function which is continuous at p? More specifically, we ask
whether there exists a (unique) function $f_p : A \to R_1$ such that

(a) $f_p(x) = Q_{f,p}(x)$ for every $x \in A - \{p\}$, and
(b) f_p is continuous at p.

The differentiability of f at p depends upon an affirmative answer to
this question.

Definition 16.1

A function $f : A \to R_1$ is said to be *differentiable at the point* $p \in A \cap A'$ iff there exists exactly one function $f_p : A \to R_1$ such that f_p is continuous at p and

$$f(x) = f(p) + (x - p)f_p(x) \qquad \text{for every } x \in A$$

The real number $f_p(p)$, called the *derivative of* f *at* p, is usually
denoted by $f'(p)$.

It is clear from this definition that differentiability is a "local"
property of a function f at a point p of its domain A. However, if f is
differentiable at p, then the real number $f_p(p)$ is uniquely determined
by p, and hence the set of all ordered pairs

$$\{\langle p, f_p(p) \rangle \mid p \in A \text{ and } f \text{ is differentiable at } p\}$$

is a subset of $A \times R_1$ which clearly satisfies the requirement of a function. Since this function is derived from f, we call it the *derivative* of f
and denote it by f'. We may now express differentiability in a
"global" form.

Definition 16.2

A function $f : A \to R_1$ is said to be differentiable (more precisely, *differentiable on A*) iff f is differentiable at p for every $p \in A$. In particular, f is differentiable on a subset H of A iff H is contained in the
domain of f'.

Note carefully that if f is differentiable on H and p, q are distinct
points of H, then $f'(p) = f_p(p)$ and $f'(q) = f_q(q)$, where f_p and f_q are distinct functions determined by Definition 16.1.

Our first theorem shows that continuity is a necessary condition
for differentiability.

Theorem 16.3

If $f : A \to R_1$ is differentiable at a point $p \in A$, then f is continuous
at p.

Proof Let f be differentiable at p. By Definition 16.1, there exists a unique function $f_p:A \to R_1$ such that f_p is continuous at p and

$$f(x) = f(p) + (x - p)f_p(x) \qquad \text{for every } x \in A$$

Since the polynomial $x - p$ and the constant $f(p)$ are also continuous at p, it follows from Theorem 9.5 that f is continuous at p. $\square$

However, continuity is not a sufficient condition for differentiability, as shown by the following example.

Example 16.4

Let $f:R_1 \to R_1$ be defined by $f(x) = |x|$. Then f is continuous on R_1 by Prob. 9.1. The student may verify (see Prob. 9.2) that $Q_{f,0}(x) = -1$ for $x < 0$ and $Q_{f,0}(x) = 1$ for $x > 0$. Clearly, there is no function f_0 such that f_0 is continuous at 0 and $f_0 = Q_{f,0}$ for $x \in R_1 - \{0\}$. Thus f is continuous at 0, but f is not differentiable at 0.

Theorem 16.5

If $f:A \to R_1$ and $g:A \to R_1$ are differentiable at a point $p \in A$, then $f + g$, fg, and $f - g$ are differentiable at p. If in addition, $g(p) \neq 0$, then f/g is differentiable at p. Furthermore, the derivatives at p satisfy

(a) $$(f + g)'(p) = f'(p) + g'(p)$$
(b) $$(fg)'(p) = f(p)g'(p) + f'(p)g(p)$$
(c) $$(f - g)'(p) = f'(p) - g'(p)$$
(d) $$\left(\frac{f}{g}\right)'(p) = \frac{f'(p)g(p) - f(p)g'(p)}{[g(p)]^2}$$

Proof Since f and g are differentiable at p, there exist unique functions $f_p:A \to R_1$ and $g_p:A \to R_1$ which are continuous at p such that for every $x \in A$,

$$f(x) = f(p) + (x - p)f_p(x)$$

and

$$g(x) = g(p) + (x - p)g_p(x)$$

Then for every $x \in A$ we have

$$\begin{aligned}
(f + g)(x) &= f(x) + g(x) \\
&= f(p) + g(p) + (x - p)[f_p(x) + g_p(x)] \\
&= (f + g)(p) + (x - p)h_p(x)
\end{aligned}$$

where $h_p(x) = f_p(x) + g_p(x)$ is clearly continuous at p.   Hence $f + g$ is differentiable at p by Definition 16.1, and

$$(f+g)'(p) = h_p(p) = f_p(p) + g_p(p) = f'(p) + g'(p)$$

which proves (a).   Also, for every $x \in A$ we have

$$\begin{aligned}
(fg)(x) &= f(x)g(x) \\
&= f(p)g(p) + (x-p)[f(p)g_p(x) \\
&\quad + f_p(x)g(p) + (x-p)f_p(x)g_p(x)] \\
&= (fg)(p) + (x-p)h_p(x)
\end{aligned}$$

where $h_p(x)$ is the expression in brackets.   It is clear that h_p is continuous at p; hence fg is differentiable at p, and

$$(fg)'(p) = h_p(p) = f(p)g_p(p) + f_p(p)g(p) = f(p)g'(p) + f'(p)g(p)$$

which proves (b).

The proof of (c) is left as an exercise.

To prove (d) we note by Prob. 15.20 that there exists a real number $\delta > 0$ such that $g(x) \neq 0$ for $x \in (p - \delta, p + \delta) \cap A$. For such x we have

$$\begin{aligned}
\frac{f}{g}(x) &= \frac{f(x)}{g(x)} = \frac{f(p)}{g(p)} + \frac{f(x)}{g(x)} - \frac{f(p)}{g(p)} \\
&= \frac{f(p)}{g(p)} + \frac{[f(p) + (x-p)f_p(x)]g(p) - f(p)[g(p) + (x-p)g_p(x)]}{g(x)g(p)} \\
&= \frac{f}{g}(p) + (x-p)h_p(x)
\end{aligned}$$

where

$$h_p(x) = \frac{f_p(x)g(p) - f(p)g_p(x)}{g(x)g(p)}$$

Clearly h_p is continuous at p, and hence f/g is differentiable at p. Furthermore,

$$\begin{aligned}
\left(\frac{f}{g}\right)'(p) = h_p(p) &= \frac{f_p(p)g(p) - f(p)g_p(p)}{[g(p)]^2} \\
&= \frac{f'(p)g(p) - f(p)g'(p)}{[g(p)]^2} \qquad \square
\end{aligned}$$

Theorem 16.6

If $f : A \to B$ is differentiable at the point $p \in A$ and $g : B \to C$ is differentiable at the point $f(p) \in B$, then $(g \circ f) : A \to C$ is differentiable at p and

$$(g \circ f)'(p) = (g' \circ f)(p) \cdot f'(p) = g'[f(p)] \cdot f'(p)$$

Proof Since f is differentiable at p, there exists a unique function $f_p: A \to B$ such that f_p is continuous at p and

$$f(x) = f(p) + (x - p)f_p(x) \qquad \text{for every } x \in A \tag{1}$$

and since g is differentiable at the point $q = f(p) \in B$, there exists a unique function $g_q: B \to C$ such that g_q is continuous at q and

$$g(y) = g(q) + (y - q)g_q(y) \qquad \text{for every } y \in B \tag{2}$$

Hence for every $x \in A$ we have

$$\begin{aligned}
(g \circ f)(x) &= g[f(x)] \\
&= g(q) + [f(x) - q]g_q[f(x)] && \text{by (2)} \\
&= g[f(p)] + [f(x) - f(p)]g_q[f(x)] \\
&= (g \circ f)(p) + (x - p)f_p(x)g_q[f(x)] && \text{by (1)} \\
&= (g \circ f)(p) + (x - p) \cdot h_p(x)
\end{aligned}$$

where $h_p(x) = (g_q \circ f)(x) \cdot f_p(x)$. The composite mapping $g_q \circ f$ is clearly continuous at p, and f_p is continuous at p; hence h_p is continuous at p. Therefore $g \circ f$ is differentiable at p, and

$$\begin{aligned}
(g \circ f)'(p) &= h_p(p) = (g_q \circ f)(p) \cdot f_p(p) \\
&= g_q[f(p)] \cdot f_p(p) = g_q(q) \cdot f_p(p) \\
&= g'(q) \cdot f'(p) = g'[f(p)] \cdot f'(p) \qquad \square
\end{aligned}$$

The proof of the following theorem is a simple exercise.

Theorem 16.7

A constant function $f: A \to \{k\}$ is differentiable at every nonisolated point of A, and its derivative is the function $f': A \cap A' \to \{0\}$.

This theorem asserts that the derivative of a constant function is the identically zero function. A very useful theorem of analysis is its converse—that a function whose derivative is identically zero is necessarily constant. We shall establish this result in two ways, first as a simple corollary of the mean value theorem and then by monotone functions.

Theorem 16.8   Mean value theorem

Let f be continuous on $[a,b]$ and differentiable on (a,b). Then there exists a point $p \in (a,b)$ such that

$$f'(p) = \frac{f(b) - f(a)}{b - a}$$

Proof   Define the function g as follows:

$$g(x) = f(a) - f(x) + \frac{f(b) - f(a)}{b - a} (x - a) \qquad \text{for } x \in [a,b]$$

It is clear that g is continuous on $[a,b]$ and differentiable on (a,b) with

$$g'(x) = -f'(x) + \frac{f(b) - f(a)}{b - a} \qquad \text{for } x \in (a,b) \tag{1}$$

Furthermore, $g(a) = g(b) = 0$. By Prob. 15.21 there exists a sequence $\langle [a_n, b_n] \rangle$ of nested closed intervals such that $g(a_n) = g(b_n)$ for each $n \in I$, and $\bigcap_{n \in I} [a_n, b_n] = \{p\}$, where $p \in (a,b)$. Since g is differentiable at p, there is a unique function g_p which is continuous at p such that

$$g(x) = g(p) + (x - p)g_p(x) \qquad \text{for every } x \in [a,b]$$

In particular, for each $n \in I$ we have

$$g(a_n) = g(p) + (a_n - p)g_p(a_n)$$

and

$$g(b_n) = g(p) + (b_n - p)g_p(b_n)$$

But $g(a_n) = g(b_n)$ for each n, and hence

$$(a_n - p)g_p(a_n) = (b_n - p)g_p(b_n) \qquad \text{for each } n \in I$$

Since $a_n - p < 0$ and $b_n - p > 0$ for each n, we see that $g_p(a_n)$ and $g_p(b_n)$ have opposite signs, so that the continuity of g_p at p implies (see Prob. 15.22) that $g'(p) = g_p(p) = 0$, and the conclusion follows from (1) above. $\square$

Corollary 16.9   Rolle's theorem

Let f be continuous on $[a,b]$ and differentiable on (a,b), and suppose that $f(a) = f(b)$. Then there exists a point $p \in (a,b)$ such that $f'(p) = 0$.

Corollary 16.10

Let f be continuous on $[a,b]$ and differentiable on (a,b), and suppose that $f'(x) = 0$ for every $x \in (a,b)$. Then f is constant on $[a,b]$.

Proof   Let x be any point of (a,b). By the mean value theorem, there exists a point $p \in (a,x)$ such that

$$f'(p) = \frac{f(x) - f(a)}{x - a}$$

Since $f'(p) = 0$, we have $f(x) = f(a)$. □

The key to our above proof of the mean value theorem is the judicious choice of the function g, and many students compare it with the magician's trick of pulling a rabbit from a hat. Actually, it is easy to see that $g = h - f$, where h is the linear function from $(a, f(a))$ to $(b, f(b))$. Our proof also depends on the nested intervals theorem, which is sometimes not available in a calculus course.

Since the primary use of the mean value theorem in calculus is to prove Corollary 16.10, we shall now establish this important corollary by another method. This different approach not only bypasses the mean value theorem but leads to an easy proof of a somewhat restricted form of that theorem. Our treatment is suggested by two articles in the May 1967 issue of *The American Mathematical Monthly*, On Being Mean to the Mean Value Theorem, by Leon Cohen, and On Avoiding the Mean Value Theorem, by Lipman Bers.

Theorem 16.11

Let f be continuous on $[a,b]$ and differentiable on (a,b), and suppose that f attains its maximum value (or its minimum value) at p, where $a < p < b$. Then $f'(p) = 0$.

Proof Let f_p be the function which is continuous at p and satisfies

$$f(x) - f(p) = (x - p)f_p(x) \qquad \text{for every } x \in (a,b)$$

Supposing f attains its maximum value at p, we have $f(x) - f(p) \leq 0$ for every $x \in (a,b)$, and hence $(x - p)f_p(x) \leq 0$ for every $x \in (a,b)$. Thus

$$f_p(x) \geq 0 \qquad \text{for } a < x < p$$

and

$$f_p(x) \leq 0 \qquad \text{for } p < x < b$$

Since f_p is continuous at p, we must have $f'(p) = f_p(p) = 0$. □

Theorem 16.12

Let f be differentiable on $H = (a,b)$. If $f'(x) > 0$ for every $x \in H$, then f is increasing on H.

Proof Assume by way of contradiction that f is not increasing on H. Then there exist points $x_1, x_2 \in H$ such that $x_1 < x_2$ and $f(x_1) \geq$

$f(x_2)$. Since f is continuous on $[x_1,x_2]$, f attains a maximum value M at some point $p \in [x_1,x_2]$. If $p \in (x_1,x_2)$, then $f'(p) = 0$ by Theorem 16.11, contradicting our hypothesis. If $f(x_2) = M$, we must also have $f(x_1) = M$, and so it suffices to consider only the case $p = x_1$. Let f_p be the function which is continuous at p and satisfies

$$f(x) = f(p) + (x - p)f_p(x) \qquad \text{for every } x \in H$$

For each x such that $p < x < x_2$, we see that $f(x) - f(p) \leq 0$, and hence $f_p(x) \leq 0$. The continuity of f_p at p then implies that $f'(p) = f_p(p) \leq 0$, again contradicting the hypothesis. □

Corollary 16.13

Let f be differentiable on $H = (a,b)$. If $f'(x) \geq 0$ for every $x \in H$, then f is nondecreasing on H.

Proof Define g by $g(x) = f(x) + \epsilon x$ for $x \in H$ and $\epsilon > 0$. Then g is differentiable on H, and $g'(x) = f'(x) + \epsilon \geq \epsilon > 0$ for every $x \in H$. By Theorem 16.12, g is increasing on H. Now assume by way of contradiction that there are points $c < d$ in H such that $f(c) > f(d)$. Choosing $\epsilon = f(c) - f(d)$, it is easy to show that $g(c) > g(d)$, which is impossible. Therefore f is nondecreasing on H. □

An alternate proof of Corollary 16.10 is now immediate, since if $f'(x) = 0$ for every $x \in H = (a,b)$, it follows from Corollary 16.13 that both f and $-f$ are nondecreasing on H, and hence f is both nondecreasing and nonincreasing on H.

Corollary 16.14

Let f be differentiable on $[a,b]$. If there is a real number M such that $f'(x) \leq M$ for every $x \in [a,b]$, then $f(b) \leq f(a) + M(b - a)$.

Proof Suppose there is such a real number M, and define $g(x) = M(x - a) - f(x)$ for each $x \in [a,b]$. Then g is differentiable, and $g'(x) = M - f'(x) \geq 0$ for every $x \in [a,b]$. By Corollary 16.13, g is nondecreasing on (a,b), so that $g(a) \leq g(b)$. The conclusion follows since $g(a) = -f(a)$ and $g(b) = M(b - a) - f(b)$. □

A differentiable function f for which f' is continuous is said to be *continuously differentiable*.

Corollary 16.15 Mean value theorem for continuously differentiable functions

If f is differentiable on $[a,b]$ and f' is continuous on $[a,b]$, then there exists a point $c \in (a,b)$ such that

$$f'(c) = \frac{f(b) - f(a)}{b - a}$$

Proof The continuous function f' attains a maximum value M and a minimum value m on the closed interval $[a,b]$ by Corollary 15.12. Let x_1 and x_2 be points of $[a,b]$ such that $f'(x_1) = M$ and $f'(x_2) = m$. Note that $-f$ is differentiable on $[a,b]$ and that $-f'$ attains the maximum value $-m$ at x_2. By Corollary 16.14 we thus have

$$f(b) \leqslant f(a) + M(b - a)$$

and

$$-f(b) \leqslant -f(a) + (-m)(b - a)$$

Combining these inequalities after multiplication by -1 in the second yields

$$f'(x_2) = m \leqslant \frac{f(b) - f(a)}{b - a} \leqslant M = f'(x_1)$$

If $m = M$, the result is trivial since f is constant on $[a,b]$; otherwise, the continuity of f' and the intermediate value theorem guarantee the existence of a point c between x_1 and x_2 (and hence in (a,b)) such that

$$f'(c) = \frac{f(b) - f(a)}{b - a} \qquad \square$$

This form of the mean value theorem is adequate for calculus and a good deal of higher analysis as well. Indeed, the requirement of continuity of the derivative is not so stringent as it might appear, since functions which are differentiable but not continuously differentiable are somewhat of a curiosity. As a matter of interest we illustrate such a function in the following example.

Example 16.16

Let $f:R_1 \to R_1$ be defined as follows:

$$f(x) = \begin{cases} x^2 \sin \dfrac{1}{x} & \text{for } x \in R_1 - \{0\} \\ 0 & \text{for } x = 0 \end{cases}$$

The student of calculus knows that f is differentiable on $R_1 - \{0\}$ and that

$$f'(x) = 2x \sin \frac{1}{x} - \cos \frac{1}{x} \qquad \text{for } x \neq 0$$

We show first that f is also differentiable at 0 and that $f'(0) = 0$. Noting that $Q_{f,0}(x) = x \sin (1/x)$ and hence that $|Q_{f,0}(x)| \leq |x|$ for $x \neq 0$, define

$$f_0(x) = \begin{cases} Q_{f,0}(x) & \text{for } x \neq 0 \\ 0 & \text{for } x = 0 \end{cases}$$

Clearly, f_0 is continuous at 0 since $|f_0(x) - f_0(0)| = |Q_{f,0}(x)| \leq |x| < \epsilon$ for all x such that $|x - 0| < \epsilon$. Hence f is differentiable at 0, and $f'(0) = f_0(0) = 0$. However, f' is not continuous at 0. To see this, define the sequence $\langle x_n \rangle$ by $x_n = 1/\pi n$ for each $n \in I$. Then $f'(x_n) = -\cos \pi n$, which is 1 or -1 according as n is odd or even. Thus the sequence $\langle x_n \rangle$ converges to 0, but the sequence $\langle f'(x_n) \rangle$ does not converge.

Another curious property of the derivative is that it preserves connected sets, regardless of whether or not it is continuous. We state this fact in the form of an intermediate value theorem.

Theorem 16.17   Intermediate value theorem for derivatives

If f is differentiable on $[a,b]$, then f' attains every value between $f'(a)$ and $f'(b)$.

Proof   We lose no generality in supposing that $f'(a) < f'(b)$; so let c be any real number such that $f'(a) < c < f'(b)$. Define the function g by

$$g(x) = f(x) - cx \qquad \text{for } x \in [a,b]$$

Clearly, g is differentiable and $g'(x) = f'(x) - c$ for $x \in [a,b]$. *Hence g* is continuous on $[a,b]$, so that g attains a minimum value at some point $p \in [a,b]$. Since $g'(a) = f'(a) - c < 0$, Prob. 16.13 asserts that g cannot attain its minimum value at a. Similarly, $g'(b) = f'(b) - c > 0$, so that g cannot attain its minimum value at b. We thus have $p \in (a,b)$. It follows from Theorem 16.11 that $g'(p) = g_p(p) = 0$ and hence $f'(p) = c$.   □

Our next result is a generalization of the mean value theorem, and the key to its proof involves pulling a much larger rabbit from the hat. In the statement of this theorem, the symbol $f^{(n)}$ represents the nth derivative of f, so that $f^{(n+1)}$ is the derivative of $f^{(n)}$. Also, $n!$ (read "n factorial") denotes the product of the first n positive integers.

Theorem 16.18 Taylor's formula with remainder

If f and its first n derivatives $f', f'', \ldots, f^{(n)}$ are continuous on $[a,b]$ and $f^{(n)}$ is differentiable on (a,b), then there exists a point $p \in (a,b)$ such that

$$f(b) = f(a) + \frac{f'(a)}{1!}(b-a) + \frac{f''(a)}{2!}(b-a)^2$$
$$+ \cdots + \frac{f^{(n)}(a)}{n!}(b-a)^n + \frac{f^{(n+1)}(p)}{(n+1)!}(b-a)^{n+1}$$

Proof We shall apply Rolle's theorem to the function F defined by

$$F(x) = f(b) - g(x) - \frac{(b-x)^{n+1}}{(b-a)^{n+1}}[f(b) - g(a)]$$

where

$$g(x) = f(x) + f'(x)(b-x) + \frac{f''(x)}{2!}(b-x)^2 + \cdots + \frac{f^{(n)}(x)}{n!}(b-x)^n$$

Let us first note that g is differentiable on (a,b), and

$$g'(x) = f'(x) + f'(x)(-1) + f''(x)(b-x)$$
$$+ f''(x)(b-x)(-1) + \frac{f'''(x)}{2!}(b-x)^2$$
$$+ \cdots + \frac{f^{(n)}(x)}{(n-1)!}(b-x)^{n-1}(-1) + \frac{f^{(n+1)}(x)}{n!}(b-x)^n$$
$$= \frac{f^{(n+1)}(x)}{n!}(b-x)^n$$

since all terms cancel except the last. Now clearly, F is continuous on $[a,b]$ and differentiable on (a,b). Furthermore, it is easy to verify that $F(a) = F(b) = 0$. Hence, by Rolle's theorem, there exists a point $p \in (a,b)$ such that $F'(p) = 0$. But

$$F'(x) = -g'(x) + \frac{(n+1)(b-x)^n}{(b-a)^{n+1}}[f(b) - g(a)]$$

and hence

$$0 = F'(p) = -\frac{f^{(n+1)}(p)}{n!}(b-p)^n + \frac{(n+1)(b-p)^n}{(b-a)^{n+1}}[f(b) - g(a)]$$

Therefore

$$f(b) - g(a) = \frac{f^{(n+1)}(p)}{(n+1)!}(b-a)^{n+1}$$

the desired conclusion. $\square$

Note that the mean value theorem is the case $n = 0$ in Taylor's formula. This important theorem has a wide variety of applications, both theoretical and practical. The following example illustrates its use in approximation theory.

Example 16.19

We shall use Taylor's formula to approximate $\sqrt{99}$ and to determine the accuracy of our approximation. Letting $f(x) = \sqrt{x}$, $a = 100$, $b = 99$, and $n = 3$, we first compute the following:

$$f(x) = x^{\frac{1}{2}} = \sqrt{x}$$

$$f'(x) = \tfrac{1}{2}x^{-\frac{1}{2}} = \frac{1}{2\sqrt{x}}$$

$$f''(x) = -\tfrac{1}{4}x^{-\frac{3}{2}} = \frac{1}{4(\sqrt{x})^3}$$

$$f'''(x) = \tfrac{3}{8}x^{-\frac{5}{2}} = \frac{3}{8(\sqrt{x})^5}$$

$$f^{(4)}(x) = -\tfrac{15}{16}x^{-\frac{7}{2}} = -\frac{15}{16(\sqrt{x})^7}$$

$$f(100) = 10$$

$$f'(100) = \tfrac{1}{20}$$

$$f''(100) = -\tfrac{1}{4,000}$$

$$f'''(100) = \tfrac{3}{800,000}$$

According to Taylor's formula, there is a real number p such that $99 < p < 100$ and

$$f(99) = f(100) + f'(100)(-1) + \frac{f''(100)}{2!}(-1)^2$$
$$+ \frac{f'''(100)}{3!}(-1)^3 + \frac{f^{(4)}(p)}{4!}(-1)^4$$
$$= 10 - \frac{1}{20} - \frac{1}{8,000} - \frac{1}{1,600,000} - \frac{5}{128(\sqrt{p})^7}$$
$$- 10 - 0.05 - 0.000125 - 0.000000625 - E$$
$$= 9.949874375 - E$$

where

$$E = \frac{5}{128(\sqrt{p}\,)^7}$$

is the error in the approximation $\sqrt{99} \approx 9.949874375$. Although we do not know the exact value of p, we can use the fact that $99 < p < 100$ to determine bounds for E. Clearly $81 < p < 100$, so that $9 < \sqrt{p} < 10$. It follows that

$$\frac{5}{128 \cdot 10^7} < E < \frac{5}{128 \cdot 9^7}$$

or in decimal equivalents, $0.0000000031 < E < 0.0000000083$. Therefore $E < 10^{-8}$, and so $\sqrt{99} = 9.9498744$ correct to seven decimal places.

PROBLEMS

16.1 Prove (c) of Theorem 16.5 using a technique similar to our proof of (a) of that theorem.

16.2 Prove Theorem 16.7.

16.3 If f is differentiable on A and k is a real number, prove that the function $g = kf$ is differentiable on A and that $g' = kf'$.

16.4 Prove (c) of Theorem 16.5 by writing $f - g = f + (-1)g$, then using Prob. 16.3 and parts (a) and (b) of Theorem 16.5.

16.5 Let $f:A \to B$, $g:B \to C$, and $h:C \to D$ be functions such that f is differentiable at p, g is differentiable at $f(p)$, and h is differentiable at $g[f(p)]$. Show that the function $h \circ g \circ f$ is differentiable at p and

$$(h \circ g \circ f)'(p) = (h' \circ g \circ f)(p) \cdot (g' \circ f)(p) \cdot f'(p)$$
$$= h'[g(f(p))] \cdot g'[f(p)] \cdot f'(p)$$

16.6 If f and g are differentiable on (a,b) and $f' = g'$ on (a,b), what can you conclude about f and g? Prove that your answer is correct.

16.7 Find a point p as asserted by the mean value theorem for the function f defined by $f(x) = x^3$ on the interval $[0,3]$.

16.8 Show that the function g defined in the proof of the mean value theorem satisfies $g = h - f$, where the graph of h is the line segment joining the points $(a,f(a))$ and $(b,f(b))$.

16.9 Prove that if f is differentiable on $H = (a,b)$ and if $f'(x) < 0$ for every $x \in H$, then f is decreasing on H.

16.10 Prove that if f is differentiable on $H = (a,b)$ and if $f'(x) \leq 0$ for every $x \in H$, then f is nonincreasing on H.

16.11 Show that the converse of Theorem 16.12 is not necessarily true by considering the function f defined by $f(x) = x^3$ on the interval $(-1,1)$.

16.12 Prove that if f is differentiable and increasing on $H = (a,b)$, then $f'(x) \geq 0$ for every $x \in H$.

16.13 Let g be differentiable on $H = [a,b]$, and suppose that g attains its minimum value on H at the point a. Prove that $g'(a) \geqslant 0$.

   Hint: The unique function g_a satisfying $g(x) - g(a) = (x - a)g_a(x)$ is continuous at a. Assuming $g_a(a) < 0$, there is a real number $\delta > 0$ such that $g_a(x) < 0$ for $a < x < a + \delta$, by Prob. 15.20. This should yield a contradiction to the fact that $g(a)$ is a minimum for g on $[a,b]$.

16.14 Let g be differentiable on $H = [a,b]$, and suppose that g attains its minimum value on H at the point b. Prove that $g'(b) \leqslant 0$.

16.15 Use the mean value theorem to show that $9.94 < \sqrt{99} < 9.95$, and compare the accuracy of this approximation with that of Example 16.19.

16.16 Use Taylor's formula with $n = 3$ to approximate $\sqrt{101}$.

16.17 Prove the *generalized law of the mean:* Let f and g be continuous on $[a,b]$ and differentiable on (a,b), with $g(a) \neq g(b)$ and $f'(x)$, $g'(x)$ not both 0 for any $x \in (a,b)$. Then there exists a point $p \in (a,b)$ such that

$$\frac{f'(p)}{g'(p)} = \frac{f(b) - f(a)}{g(b) - g(a)}$$

Hint: Apply Rolle's theorem to the function h, where

$$h(x) = f(x) - f(a) - \frac{f(b) - f(a)}{g(b) - g(a)} \, [g(x) - g(a)]$$

16.18 Show that the mean value theorem is the special case of the generalized law of the mean (see Prob. 16.17), where g is the identity function on $[a,b]$.

16.19 Prove Taylor's formula (Theorem 16.18) by applying the generalized law of the mean to the functions F and G defined by

$$F(x) = f(b) - f(x) - f'(x)(b - x) - \cdots - \frac{f^{(n)}(x)}{n!} \, (b - x)^n$$

$$G(x) = \frac{(b - x)^{n+1}}{(n + 1)!}$$

17
The Cantor Set and the Cantor Function

This chapter is devoted to the study of an important set of real numbers known as the *Cantor set*. Because of its many peculiar properties, the Cantor set enjoys a prominent role in analysis, often serving as a counterexample for many of our intuitive ideas. Since our work with this set will require some facility with binary and ternary expansions of real numbers, we begin with a short discussion of such expansions.

Perhaps the easiest way to understand binary and ternary expansions is to review the fundamentals of *decimal* expansions of real numbers. Since the decimal system is based on a scale of 10, we have exactly ten distinct digits $0, 1, 2, \ldots, 9$ which serve as coefficients of integral powers of 10 in the decimal expansion. For example, the real number 20,742.305 represents the sum given by

$$2 \cdot 10^4 + 0 \cdot 10^3 + 7 \cdot 10^2 + 4 \cdot 10^1 + 2 \cdot 10^0$$
$$+ 3 \cdot 10^{-1} + 0 \cdot 10^{-2} + 5 \cdot 10^{-3}$$

which may also be written as

$$20{,}000 + 700 + 40 + 2 + \tfrac{3}{10} + \tfrac{5}{1000}$$

In particular, if b is a real number such that $0 \leq b \leq 1$, then the decimal expansion of b has the form

$$0.b_1 b_2 b_3 b_4 \cdots \tag{1}$$

where each b_i, $i \in I$, is one of the digits $0, 1, 2, \ldots, 9$. Expression (1) represents the sum

$$0 + b_1 \cdot 10^{-1} + b_2 \cdot 10^{-2} + b_3 \cdot 10^{-3} + b_4 \cdot 10^{-4} + \cdots$$

which we may write as

$$\frac{b_1}{10} + \frac{b_2}{10^2} + \frac{b_3}{10^3} + \frac{b_4}{10^4} + \cdots \tag{2}$$

Note that if each $b_i = 9$, then (2) is a geometric series with first term $a = \frac{9}{10}$ and common ratio $r = \frac{1}{10}$. Hence the series converges, by Theorem 13.6, and its sum is $S = \frac{9}{10}/(1 - \frac{1}{10}) = 1$. Since (1) and (2) are equivalent expressions for the decimal expansion of a real number, it follows that

$$0.9999 \cdots = 1.0000 \cdots$$

A slight modification of this argument can be used to show that any real number b of the form (1) which eventually ends in 9s, that is $b_i = 9$ for every $i \geq k$, for some $k \in I$, has an equivalent decimal expansion which eventually ends in 0s. Furthermore, it can be shown that a real number b of the form (1) has two such equivalent decimal expansions iff $b = m/10^k$ for some nonnegative integer k, where m is an integer such that $0 \leq m \leq 10^k$. Thus b has either a unique decimal expansion or exactly two equivalent decimal expansions, one eventually ending in 9s and the other eventually ending in 0s.

Although this discussion of decimal expansions is sketchy, it does exhibit the basic pattern for describing other expansions of real numbers. The *ternary* system is based on a scale of 3, and hence we have exactly three distinct digits 0, 1, 2 which serve as coefficients of integral powers of 3 in the ternary expansion. We indicate the scale by enclosing it in brackets after the expansion. For example, the real number

$$12012.112 \quad [3]$$

represents the sum given by

$$1 \cdot 3^4 + 2 \cdot 3^3 + 0 \cdot 3^2 + 1 \cdot 3^1 + 2 \cdot 3^0 + 1 \cdot 3^{-1} + 1 \cdot 3^{-2} + 2 \cdot 3^{-3}$$

In particular, if b is a real number such that $0 \leq b \leq 1$, then the ternary expansion of b has the form

$$0.b_1 b_2 b_3 \cdots \quad [3] \tag{3}$$

where each $b_1, i \in I$, is one of the digits 0, 1, 2. Expression (3) represents the sum

$$0 + b_1 \cdot 3^{-1} + b_2 \cdot 3^{-2} + b_3 \cdot 3^{-3} + \cdots$$

which we may write as

$$\frac{b_1}{3} + \frac{b_2}{3^2} + \frac{b_3}{3^3} + \frac{b_4}{3^4} + \cdots \tag{4}$$

Note that if each $b_i = 2$, then (4) is a geometric series with first term $a = \frac{2}{3}$ and common ratio $r = \frac{1}{3}$. Hence the series converges, by Theorem 13.6, and its sum is $S = \frac{2}{3}/(1 - \frac{1}{3}) = 1$. Since (3) and (4) are equivalent expressions for the ternary expansion of a real number, it follows that

$$0.2222 \cdots = 1.0000 \cdots \qquad [3]$$

Again, a slight modification of this argument can be used to show that any real number of the form (3) which eventually ends in 2s has an equivalent ternary expansion which eventually ends in 0s. It can also be shown that a real number b of the form (3) has two such equivalent ternary expansions iff $b = m/3^k$ for some nonnegative integer k, where m is an integer such that $0 \leqslant m \leqslant 3^k$. Thus b has either a unique ternary expansion or exactly two equivalent ternary expansions, one eventually ending in 2s and the other eventually ending in 0s.

The *binary* system is based on a scale of 2, and hence there are exactly two distinct digits 0, 1 which serve as coefficients of integral powers of 2 in the binary expansion of a real number. The details of this system are left as an exercise.

With this background, we may now proceed with our discussion of the Cantor set. Throughout this section, the symbol J will be used to represent the closed unit interval $[0,1]$, and we begin by defining a certain subset G of the set J:

$$G = \{x \in J \mid \text{the ternary expansion}$$
$$\text{for } x \textit{ must } \text{contain at least one } 1\} \tag{5}$$

The set G is well defined for any x in J which has a unique ternary expansion but requires some clarification for the members of J which have two equivalent ternary expansions. If either of these two equivalent ternary expansions of x contains no 1, then it follows from (5) that $x \notin G$. For example, the real number

$$\tfrac{1}{3} = 0.10000 \cdots \qquad [3]$$

does not belong to G since it has an equivalent ternary expansion

$$\tfrac{1}{3} = 0.02222 \cdots \qquad [3]$$

Similarly, since the real number

$$0.12222 \cdots = 0.20000 \cdots \qquad [3]$$

we conclude that $\frac{2}{3} \notin G$.

It should be evident from the above discussion that if the real number x in J has a ternary expansion containing precisely one 1 followed immediately by all 2s or followed immediately by all 0s, then x has an equivalent ternary expansion containing no 1, and hence $x \notin G$. We shall use this fact in some of our proofs, beginning with the following fundamental lemma.

Lemma 17.1

A point x of J is in G iff there exist positive integers k and m satisfying the following conditions:

(a) $$0 < m < m + 1 < 3^k$$

(b) $$\frac{m}{3^k} < x < \frac{m+1}{3^k}$$

(c) $$m = 3t + 1 \qquad \text{for some nonnegative integer } t$$

Proof Suppose first that $x \in G$ and that a ternary expansion of x is given by (3). Let k be the smallest positive integer such that $b_k = 1$. Then we have

$$x = 0.b_1 b_2 \cdots b_{k-1} 1 b_{k+1} b_{k+2} \cdots \qquad [3]$$

where $b_i \neq 1$ for $i = 1, 2, \ldots, k - 1$. It follows from the definition of G and our discussion above that the digit $b_k = 1$ cannot be followed immediately by all 2s or by all 0s. Thus it is evident that

$$0.b_1 b_2 \cdots b_{k-1} 1 < x < 0.b_1 b_2 \cdots b_{k-1} 2 \qquad [3] \qquad (6)$$

Note that

$$0.b_1 b_2 \cdots b_{k-1} = \frac{b_1}{3} + \frac{b_2}{3^2} + \cdots + \frac{b_{k-1}}{3^{k-1}}$$

$$= \frac{1}{3^{k-1}} (b_1 \cdot 3^{k-2} + b_2 \cdot 3^{k-3} + \cdots + b_{k-2} \cdot 3 + b_{k-1})$$

If we denote by t the nonnegative integer represented by the sum in parentheses, then (6) becomes

$$\frac{t}{3^{k-1}} + \frac{1}{3^k} < x < \frac{t}{3^{k-1}} + \frac{2}{3^k}$$

or

$$\frac{3t+1}{3^k} < x < \frac{3t+2}{3^k}$$

Defining $m = 3t + 1$, we see that (a) and (b) are satisfied.

Conversely, suppose that x is a point of J and k, m positive integers such that (a), (b), and (c) are satisfied. We must show that $x \in G$. From (b) and (c) we have

$$\frac{m}{3^k} = \frac{3t+1}{3^k} = \frac{t}{3^{k-1}} + \frac{1}{3^k}$$

and hence $m/3^k$ has a ternary expansion of the form

$$\frac{m}{3^k} = 0.a_1 a_2 \cdots a_{k-1} 10000 \cdots \qquad [3]$$

Similar reasoning shows that $(m + 1)/3^k$ has a ternary expansion with the same digits except that $a_k = 2$ instead of 1. However, we decide to use the equivalent ternary expansion ending in 2s, so that

$$\frac{m+1}{3^k} = 0.a_1 a_2 \cdots a_{k-1} 12222 \cdots \qquad [3]$$

It then follows from (b) that

$$x = 0.a_1 a_2 \cdots a_{k-1} 1 a_{k+1} a_{k+2} \cdots \qquad [3]$$

where at least one $a_i \neq 0$ for $i > k$ and at least one $a_i \neq 2$ for $i > k$. Thus the ternary expansion of x must contain at least one 1, and so $x \in G$. $\square$

As an immediate application of this lemma, we have the following theorem.

Theorem 17.2

(a) The set G is open.

(b) If p and q are distinct points of J, then there is a point x of G between p and q.

(c) $\overline{G} = J$.

Proof To prove (a), let x be any point of G. By Lemma 17.1, there exist positive integers k and m satisfying the three conditions of the lemma. But (b) of the lemma asserts that every point of J between $m/3^k$ and $(m + 1)/3^k$ belongs to G. Thus we have

$$x \in \left(\frac{m}{3^k}, \frac{m+1}{3^k} \right) \subset G$$

To prove (b), we may suppose without loss of generality that $p < q$, so let us define $r = q - p > 0$. Since $\lim (1/3^n) = 0$, there exists a positive integer k such that $1/3^k < r/4$. Let y be the smallest positive integer such that $y/3^k > p$. Then $(y - 1)/3^k \leqslant p$, and hence

$$q = p + r > p + \frac{4}{3^k} = \frac{p \cdot 3^k + 4}{3^k} \geqslant \frac{y - 1 + 4}{3^k} = \frac{y + 3}{3^k}$$

Thus the four points $y/3^k$, $(y + 1)/3^k$, $(y + 2)/3^k$, $(y + 3)/3^k$ are contained in the open interval (p,q). Since every third successive positive integer is divisible by 3, it follows at once that one of the three integers y, $y + 1$, $y + 2$ must have the form $3t + 1$ for some nonnegative integer t. The result then follows from Lemma 17.1.

To prove (c), it suffices to show that $J \subset \bar{G}$ (why?); so let p be any point of J. Since we want to show that $p \in \bar{G} = G \cup G'$, let us suppose that $p \notin G$. Let (r,p,s) be any deleted open interval about p. Since $p \in J = [0,1]$, we see that $(r,p,s) \cap J \neq \varnothing$, and so let q be any point in this intersection. Then p and q are distinct points of J; hence, by (b) above, there is a point x of G between p and q, and clearly $x \in (r,p,s)$. Hence by Definition 5.10, we have $p \in G'$.   □

The relationship $\bar{G} = J$ asserted in Theorem 17.2 says that G is dense in J, according to the following definition.

Definition   17.3

A set A is said to be *dense* in B iff $A \subset B \subset \bar{A}$.

We remark that if the set B in Definition 17.3 is closed, then A is dense in B iff $\bar{A} = B$. In particular, a subset A of R_1 is dense in R_1 iff $\bar{A} = R_1$, and in this case, we sometimes say that A is *everywhere dense*. The student is requested in the problems to prove that the set of all rational numbers is dense in R_1.

Definition   17.4

The *Cantor set* is the subset of $J = [0,1]$ consisting of precisely those real numbers which have ternary expansions containing no 1s.

Theorem   17.5

(a)   The Cantor set is the set $J - G$.
(b)   The Cantor set is closed.

(c) The Cantor set contains no connected subset consisting of more
than one point.

(d) The Cantor set is uncountable.

Proof (a) is immediate from the definition of G and Definition 17.4.
Part (b) follows at once from (a) above and Theorem 17.2a, since
the complement of the Cantor set is the union of the three open
sets G, $\{x \mid x < 0\}$, $\{x \mid x > 1\}$ and hence is open. Part (c) is a corollary to Theorem 17.2b.

To prove (d), we shall define a mapping from the Cantor set
onto the uncountable set J. It will then follow from the contrapositive of Theorem 11.2 that the Cantor set is uncountable.
Let us denote the Cantor set by K, so that each point x of K has a
ternary expansion of the form

$$x = 0.a_1a_2a_3 \cdots \quad [3]$$

where for each $i \in I$, a_i is either 0 or 2. For each $i \in I$, we
define $b_i = a_i/2$, so that each b_i is either 0 or 1, and let

$$y = 0.b_1b_2b_3 \cdots \quad [2] \tag{7}$$

Finally, for each x in K, define

$$f(x) = y$$

where y is determined from x by (7). Since every real number in
the closed interval [0,1] has a binary expansion of the form (7), it
is clear that f maps K onto J, thus completing the proof. □

An interesting exercise (see Prob. 17.14) is the evaluation of f at
various points of K.

Our investigation of the Cantor set thus far has proceeded by
strictly analytical methods and has provided a considerable amount of
information about this interesting set. Buried in these results, however, is another method of describing the Cantor set, a method which
may lead to a keener insight of the set's structure by appealing to our
intuition.

Let us return for a moment to Lemma 17.1 and suppose we
choose $k = 1$. Then the only positive integer m satisfying (a) and (c)
of the lemma is $m = 1$, and it follows from (b) that every real number x
such that $\frac{1}{3} < x < \frac{2}{3}$ is a member of the set G. Thus, as a first step in
constructing the Cantor set, we may *remove* the open middle third of
the closed unit interval $J = [0,1]$. If we now let $k = 2$ in Lemma 17.1,
the only positive integers m satisfying both (a) and (c) of the lemma
are $m = 1$, 4, 7. Then according to (b), the open intervals
$(\frac{1}{9},\frac{2}{9})$, $(\frac{4}{9},\frac{5}{9})$, $(\frac{7}{9},\frac{8}{9})$ belong to G. Of course, the open interval $(\frac{4}{9},\frac{5}{9})$ is con-

tained in the open interval $(\frac{1}{3},\frac{2}{3})$, which has already been removed at the first step, so we may ignore it here. Note that $(\frac{1}{9},\frac{2}{9})$ is the open middle third of the closed interval $[0,\frac{1}{3}]$ and $(\frac{7}{9},\frac{8}{9})$ is the open middle third of the closed interval $[\frac{2}{3},1]$. Thus, as a second step in constructing the Cantor set, we may *remove* the open middle third of each of the two closed subintervals of J determined by the first step. The student may verify that the third step in constructing the Cantor set (achieved by letting $k = 3$ in Lemma 17.1) consists of the removal of the open middle third of each of the four closed subintervals $[0,\frac{1}{9}]$, $[\frac{2}{9},\frac{1}{3}]$, $[\frac{2}{3},\frac{7}{9}]$, $[\frac{8}{9},1]$ of J determined by the second step. This process is continued for each $k \in I$, where each step removes the open middle third of each of the closed subintervals of J determined by the previous step. All the points which are removed constitute the set G, and the points of J which remain are precisely the Cantor set.

It should be intuitively evident from the above discussion that G is the union of a collection of open intervals and hence that G is open, a fact already established by Theorem 17.2a. However, considerably more can be said about the set G, and since it will be useful in the proof of our next major theorem, we state it here as a lemma.

Lemma   17.6

The set G is the union of a countable collection of disjoint open intervals the sum of whose lengths is 1.

Proof   We leave as an exercise the proof that G is the union of a countable collection $\{J_i\}$ of disjoint open intervals. We shall prove that the sum of the lengths of the J_i is 1. According to our discussion above and Lemma 17.1, G contains the open interval $(\frac{1}{3},\frac{2}{3})$ of length $\frac{1}{3}$, and for each integer $k \geq 2$, G contains 2^{k-1} open intervals each of length $1/3^k$. If we let L denote the sum of the lengths of all these open intervals, it follows that

$$L = \frac{1}{3} + 2\frac{1}{3^2} + 4\frac{1}{3^3} + 8\frac{1}{3^4} + \cdots$$

$$= \frac{1}{3} + \frac{2}{3^2} + \frac{2^2}{3^3} + \frac{2^3}{3^4} + \cdots \tag{8}$$

This last expression is a geometric series with first term $a = \frac{1}{3}$ and common ratio $r = \frac{2}{3}$. By Theorem 13.6, the series converges, and its sum is $L = 1$.   □

Since the Cantor set is the set $J - G$, it is what remains after G is removed from the closed interval $[0,1]$. Clearly, the end points of all the removed open intervals must then be members of the Cantor set, as well as any cluster points of the set of end points (since the Cantor set is closed). Actually, our next theorem will display the remarkable property that every point of the Cantor set is a cluster point of the set. Closed sets which enjoy this property are given a special name, according to the following definition.

Definition   17.7

A set H in R_1 is called *perfect* iff $H = H'$.

We have encountered a few perfect sets in our work thus far; namely, the empty set $\varnothing$, the set R_1, closed intervals, and closed rays.

One additional set property of importance in analysis will be introduced here. Recall that a set A is everywhere dense iff $\bar{A} = R_1$; consequently, A is everywhere dense iff every open interval is contained in the closure of A. Now suppose we consider the opposite extreme where the closure of A contains no open interval. It seems reasonable in this case to say that A is nowhere dense, and this is precisely what we do.

Definition   17.8

A set H is said to be *nowhere dense* iff $\bar{H}$ contains no open interval.

It is easy to see that every uniformly isolated set, and hence every finite set, is nowhere dense. Another example is the set $\{1/n \mid n \in I\}$ or for that matter, the range of any convergent sequence together with its limit. The ease of finding these examples rests primarily on the fact that for any such set H, its derived set H' is either the empty set or a singleton. The task of finding a nowhere dense set H becomes much more difficult if we insist that H' be infinite. In fact, let us tease our intuition by posing the following question: Is it possible for a nonempty perfect set to be nowhere dense? Our next theorem answers this question affirmatively, thus demonstrating some of the surprising properties of the Cantor set.

Theorem   17.9

The Cantor set is
(a)   Nowhere dense
(b)   Perfect
(c)   Of measure zero

Proof   (*a*) follows immediately from Theorem 17.5*b* and *c*.

To prove (b), let p be any point of the Cantor set. Then p has a ternary expansion of the form

$$p = 0.a_1 a_2 a_3 \cdots \qquad [3]$$

where each a_i, $i \in I$, is either 0 or 2. For any $\epsilon > 0$, choose a positive integer k such that $1/3^k < \epsilon$. If $a_{k+1} = 0$, define $b_{k+1} = 2$, and if $a_{k+1} = 2$, define $b_{k+1} = 0$. Now define q to be the real number whose ternary expansion is identical to that of p except that b_{k+1} has replaced a_{k+1}. Clearly, q is a point of the Cantor set, distinct from p, and

$$0 < |p - q| = \frac{|a_{k+1} - b_{k+1}|}{3^{k+1}} = \frac{2}{3^{k+1}} < \frac{1}{3^k} < \epsilon$$

Thus for any $\epsilon > 0$, the deleted open interval $(p - \epsilon, p, p + \epsilon)$ contains a point of the Cantor set. Hence p is a cluster point of the Cantor set K, and we have $K \subset K' \subset \bar{K} = K$. Therefore $K = K'$.

To prove (c), let $\epsilon > 0$ be given, and note from Lemma 17.6 that G is the union of a countable collection $\{J_i\}$ of disjoint open intervals the sum of whose lengths is 1. For each positive integer n, let S_n be the partial sum of the first n terms of the series (8). Since $\langle S_n \rangle$ is an increasing sequence converging to 1, there exists a positive integer k such that $S_k > 1 - \epsilon/3$. Thus there are a finite number $J_1, J_2, \ldots, J_N$ of disjoint open intervals in G, the sum of whose lengths is greater than $1 - \epsilon/3$. We suppose these intervals numbered from left to right, so that

$$J_i = (a_i, b_i) \qquad \text{for } 1 \leq i \leq N$$

where $0 < a_1 < b_1 < a_2 < b_2 < \ldots < a_N < b_N < 1$. Then the Cantor set is contained in the union of the closed intervals

$$[0, a_1] \cup [b_1, a_2] \cup [b_2, a_3] \cup \cdots \cup [b_N, 1]$$

and the sum S of the lengths of these intervals is less than $\epsilon/3$. Let $b_0 = 0$, $a_{N+1} = 1$, and for each $i = 1, 2, \ldots, N + 1$ define the open interval G_i by

$$G_i = \left(b_{i-1} - \frac{\epsilon}{6N}, \ a_i + \frac{\epsilon}{6N} \right)$$

Then $\{G_i\}$ is a countable collection of open intervals covering the Cantor set. Furthermore, the sum of the lengths of G_1, $G_2, \ldots, G_{N+1}$ is

$$S + (N + 1) \cdot \frac{\epsilon}{3N} \leq S + (N + N) \cdot \frac{\epsilon}{3N} < \frac{\epsilon}{3} + \frac{2\epsilon}{3} = \epsilon \qquad \square$$

Before discussing the Cantor function, we first develop some notation and general results for monotone functions.

Definition   17.10

Let f be defined and nondecreasing on J. Then we write

$$f(c+) = \inf \{f(x) \mid x \in J \text{ and } x > c\} \qquad \text{for each } c \in [0,1)$$
$$f(c-) = \sup \{f(x) \mid x \in J \text{ and } x < c\} \qquad \text{for each } c \in (0,1]$$

The real number $f(c+)$ is called the *right-hand limit of f at c*, and $f(c-)$ is called the *left-hand limit of f at c*.

Note that we have not defined a left-hand limit of f at 0 or a right-hand limit of f at 1. However, $f(0+)$ exists and satisfies $f(0) \leqslant f(0+)$; similarly, $f(1-)$ exists and satisfies $f(1-) \leqslant f(1)$. Furthermore, for any $c \in (0,1)$, both $f(c-)$ and $f(c+)$ exist and satisfy

$$f(c-) \leqslant f(c) \leqslant f(c+)$$

The reader should verify the above assertions, which are easy consequences of the monotonicity of f plus the least upper bound axiom (or the greatest lower bound theorem).

Definition   17.11

Let f be defined and nondecreasing on J. The real number

$$j_f(c) = f(c+) - f(c-) \qquad \text{for } c \in (0,1)$$

is called the *jump* of f at the point c. At the end points of J the jump of f is given by

$$j_f(0) = f(0+) - f(0) \qquad \text{and} \qquad j_f(1) = f(1) - f(1-)$$

We clearly have $0 \leqslant j_f(c) \leqslant f(1) - f(0)$ for each $c \in J$, and when $j_f(c) > 0$, we say that f has a *finite jump discontinuity at c*. The following theorem is self-evident, and its easy proof is left as an exercise.

Theorem   17.12

Let f be defined and nondecreasing on J. Then for any $c \in J$,

(a) f is continuous from the left at c iff $f(c-) = f(c)$.
(b) f is continuous from the right at c iff $f(c+) = f(c)$.
(c) f is continuous at c iff $j_f(c) = 0$.

Corollary   17.13

Let f be defined and nondecreasing on J. Then the only discontinuities of f are finite jump discontinuities.

Corollary   17.14

Let f be defined and nondecreasing on J. Then f has at most a countable number of points of discontinuity.

The way is now cleared to discuss the Cantor function, and we begin with the mapping $f : K \xrightarrow{\text{onto}} J$ defined in the proof of Theorem 17.5d. The idea is to extend the domain of f to all of J, and this is easily done using Prob. 17.14. We see that the points $\frac{1}{3}$ and $\frac{2}{3}$ are in K, with $f(\frac{1}{3}) = f(\frac{2}{3}) = \frac{1}{2}$, whereas the open interval $(\frac{1}{3}, \frac{2}{3}) \subset G$. Define

$$\phi(x) = \frac{1}{2} \qquad \text{for} \qquad \frac{1}{3} \leqslant x \leqslant \frac{2}{3}$$

Now consider the points $\frac{1}{9}, \frac{2}{9} \in K$, with $f(\frac{1}{9}) = f(\frac{2}{9}) = \frac{1}{4}$, and $(\frac{1}{9}, \frac{2}{9}) \subset G$. Define

$$\phi(x) = \tfrac{1}{4} \qquad \text{for} \qquad \tfrac{1}{9} \leqslant x \leqslant \tfrac{2}{9}$$

For the points $\frac{7}{9}, \frac{8}{9} \in K$, with, $f(\frac{7}{9}) = f(\frac{8}{9}) = \frac{3}{4}$, and $(\frac{7}{9}, \frac{8}{9}) \subset G$, define

$$\phi(x) = \tfrac{3}{4} \qquad \text{for} \qquad \tfrac{7}{9} \leqslant x \leqslant \tfrac{8}{9}$$

Clearly we can continue this process, defining ϕ on each open middle third (a,b) (removed in the construction of K) to be the common value of f at the end points a and b (which belong to K). Thus ϕ is defined on all of J, and $\phi(x) = f(x)$ for $x \in K$.

The mapping ϕ is called the *Cantor function*. It has some interesting properties, which we list as a theorem.

Theorem   17.15

The Cantor function ϕ has the following properties:

(a) ϕ is defined on all of J.
(b) ϕ maps J onto J, with $\phi(0) = 0$ and $\phi(1) = 1$.
(c) ϕ is nondecreasing on J.
(d) ϕ is continuous on J.
(e) ϕ is differentiable on G, and $\phi'(x) = 0$ for each $x \in G$.

Proof   (a), (b), and (c) are immediate from the definition of ϕ. To prove (d), note that (b) and (c) imply that ϕ has no jump discontinuities. Then (a) and (c) together with Corollary 17.13 assert that

ϕ is continuous. For (e), let x be any point of G. Since G is open, there exists an open interval $A = (x - \delta, x + \delta)$ such that $x \in A \subset G$. It follows from the definition of ϕ that ϕ is constant on A. Hence ϕ is differentiable at x, and $\phi'(x) = 0$. $\square$

We say that a property P holds "almost everywhere" in a set H if the subset of H for which P fails is a set of measure zero. Using this terminology and the fact that the Cantor set K is a set of measure zero, we may rephrase (e) above as follows: The derivative of ϕ vanishes almost everywhere in J. The Cantor function is thus a nice example of a continuous nonconstant monotone function whose derivative vanishes almost everywhere.

PROBLEMS

17.1   Let b be a real number such that $0 < b < 1$, and suppose that b has a decimal expansion of the form

$$b = 0.b_1 b_2 \cdots b_k 9999 \cdots$$

where $b_k \neq 9$. Show that b has an equivalent decimal expansion which eventually ends in 0s.

17.2   Let b be a real number such that $0 < b < 1$ and suppose that b has a ternary expansion of the form

$$b = 0.b_1 b_2 \cdots b_k 2222 \cdots \qquad [3]$$

where $b_k \neq 2$. Show that b has an equivalent ternary expansion which eventually ends in 0s.

17.3   Using our discussions of the decimal and ternary expansions as a model, write a brief description of the binary system.

17.4   Prove that a set A is dense in B iff every open interval about each point of B contains at least one point of A.

17.5   Prove that the rational numbers are dense in R_1.

17.6   Complete the proof of Lemma 17.6.

17.7   Prove that the union of a finite number of closed intervals is a perfect set.

17.8   Prove that every perfect set is closed.

17.9   Prove that the union of a finite number of perfect sets is a perfect set.

17.10   State carefully what is meant by the assertion that a set is *not* everywhere dense.

17.11   State carefully what is meant by the assertion that a set is *not* nowhere dense.

17.12   Give an example of a set which is neither everywhere dense nor nowhere dense.

17.13   Prove that a closed set is nowhere dense iff its complement is everywhere dense.

17.14 For the mapping $f: K \xrightarrow{\text{onto}} J$ defined in the proof of Theorem 17.5d, verify the following:

(a) $$f(\tfrac{1}{3}) = f(\tfrac{2}{3}) = \tfrac{1}{2}$$

(b) $$f(\tfrac{1}{9}) = f(\tfrac{2}{9}) = \tfrac{1}{4} \qquad f(\tfrac{7}{9}) = f(\tfrac{8}{9}) = \tfrac{3}{4}$$

(c) $$f(\tfrac{1}{27}) = f(\tfrac{2}{27}) = \tfrac{1}{8} \qquad f(\tfrac{7}{27}) = f(\tfrac{8}{27}) = \tfrac{3}{8}$$
$$f(\tfrac{19}{27}) = f(\tfrac{20}{27}) = \tfrac{5}{8} \qquad f(\tfrac{25}{27}) = f(\tfrac{26}{27}) = \tfrac{7}{8}$$

17.15 Let f be defined and nondecreasing on $J = [0,1]$. Show that:
(a) $f(0+)$ exists, and $f(0) \leqslant f(0+)$.
(b) $f(1-)$ exists, and $f(1-) \leqslant f(1)$.
(c) For any $c \in (0,1)$, both $f(c-)$ and $f(c+)$ exist, and $f(c-) \leqslant f(c) \leqslant f(c+)$.

17.16 Prove Theorem 17.12.

17.17 Prove Corollary 17.14. *Hint:* Use Corollary 17.13 and the fact that each finite jump discontinuity yields an open interval in the range of f.

17.18 Let f be defined and nondecreasing on $J = [0,1]$, and let $M = f(1) - f(0)$. For each $n \in I$, define

$$H_n = \left\{ x \in J \mid j_f(x) \geqslant \frac{M}{n} \right\}$$

(a) Show that each H_n is a finite set.
(b) Devise an alternate proof of Corollary 17.14.

Index

Index